Encyclopedia of Global Warming and Environmental Ecosystem

Encyclopedia of Global Warming and Environmental Ecosystem

Dr. R. Kailash

AMIGA PRESS INC.

DELHI (INDIA)

AMIGA PRESS INC.
H.O.: 28-A, Indra Park, Chander Nagar
Delhi-110051 (India)
Tel.: 011-22504028, +91-98736-57000
E-mail: amigapress@gmail.com
amigaeditorial@gmail.com
Website : www.amigapress.co.in

This edition is published by Amiga Press Inc.

ISBN: 978-93-845-33-397

Price : ₹ 1950/-

First Published 2018

Printed at : AMIGA PRESS INC

Preface

Global warming is one of the most serious threats that the planet Earth and its inhabitants may ever face. The Earth's climate has changed throughout the Earth's history--from glacial periods when ice covered significant portions of the Earth to interglacial periods when ice either retreated to poles or melted entirely.

The planet is warming, from North Pole to South Pole, and everywhere in between. Globally, the mercury is already up more than 1 degree Fahrenheit (0.8 degree Celsius), and even more in sensitive polar regions. And the effects of rising temperatures aren't waiting for some far-flung future. They're happening right now. Signs are appearing all over, and some of them are surprising. The heat is not only melting glaciers and sea ice, it's also shifting precipitation patterns and setting animals on the move.

Although the science remains uncertain, 1990 marked the year when many governments accepted the impending reality of global climate change. Some governments began to announce targets for reductions in carbon dioxide in order to prompt the international negotiation process. Some governments remain unconvinced that the greenhouse effect is real. Others accept the science, but not the need to act dramatically. This volume is concerned not with the scientific debate as such, but with the basis for climate change policy in OECD countries. Crucial to the whole process of policy development is an assessment of the impacts of climate change.

These impacts will appear first as physical changes in the environment, but will eventually translate into social and economic changes as well. Impacts are important, essentially because they represent the potential benefits that might be realised from taking political action against the climate change problem.

Global climate change has already had observable effects on the environment. Glaciers have shrunk, ice on rivers and lakes is breaking up earlier, plant and animal ranges have shifted and trees are flowering sooner.

This book is a highly informative source on global warming and climate change issues that are perhaps the greatest threat to this planet. It will be a valuable reference tool for environmental scientists, activists and scholars in the field.

–Author

Contents

Preface 5

1. **Introduction to Global Warming** 9

   Global Warming: Causes and Resulting Climate Change; Concentration of Greenhouse Gases in the Atmosphere; Atmospheric Aerosols; Evidence for Global Warming; Detection of Anthropogenic Global Warming; Greenhouse Modelling versus Observation

2. **Limits of Global Warming and Environmental Management** 40

   Global Warming Scenarios; Environmental Management; Broadening Environmental Management

3. **Effects of Global Warming on Ecosystems** 89

   Results of Global Warming on Ecosystems; Observed and expected effects on Ecosystems; Ecosystem Services in Global Warming; Basic Components of the Ecosystem; Global Warming Stress to Coral Reefs; The Effects of Global Warming on Open Oceans; The Effects of Global Warming on Coastal Locations

4. **Ozone Depletion** 123

   What Causes Ozone Depletion?; Origin of Ozone; Oxygen and Ozone; Other Effects of Ozone Depletion; Observed Changes in Stratospheric Ozone; Ozone Depleting Substances

5. **Carbon Dioxide in Earth's Atmosphere** 150

   Current Concentration; Composition of Atmosphere; Percentage Composition of Pure Dry Air; Carbon Monoxide in Nature; Forms of Oxygen and their Distribution; Earth's Atmosphere

6. **Climate Change** **186**

Terminology; Astronomical Theory and Abrupt Climate Changes; Biogeochemistry and Radiative Forcing; Climate Change Mitigation; Climate Change and Ecosystems; Time Scale of Climatic Change; The Effects of Climate Change on Ecosystems; Impact of Climate Change on Agriculture; Physical Evidence for Climatic Change; Measurement of Climate Elements; Changes in Glaciers, Ice Sheets and Ice Shelves; Radiative Forcing and Climate Sensitivity

1

Introduction to Global Warming

Measurements of temperature taken by instruments all over the world, on land and at sea have revealed that during the 20th century the Earth's surface and lowest part of the atmosphere warmed up on average by about 0.6°C. During this period, man-made emissions of greenhouse gases, including carbon dioxide, methane and nitrous oxide have increased, largely as a result of the burning of fossil fuels for energy and transportation, and land use changes including deforestation for agriculture. In the last 20 years, concern has grown that these two phenomena are, at least in part, associated with each other. That is to say, global warming is now considered most probably to be due to the increases in greenhouse gas emissions and concurrent increases in atmospheric greenhouse gas concentrations, which have enhanced the Earth's natural greenhouse effect. Whilst other natural causes of climate change can cause global climate to change over similar periods of time, computer models demonstrate that in all probability there is a real discernible human influence on the global climate.

If the climate changes as current computer models have projected, global average surface temperature could be anywhere from 1.4 to 5.8°C higher by the end of the 21st century than in 1990. To put this temperature change into context, the increase in global average surface temperature which brought the Earth out of the last major ice age 14,000 years ago was of the order of 4 to

5°C. Such a rapid change in climate will probably be too great to allow many ecosystems to suitably adapt, and the rate of species extinction will most likely increase. In addition to impacts on wildlife and species biodiversity, human agriculture, forestry, water resources and health will all be affected. Such impacts will be related to changes in precipitation (rainfall and snowfall), sea level, and the frequency and intensity of extreme weather events, resulting from global warming. It is expected that the societies currently experiencing existing social, economic and climatic stresses will be both worst affected and least able to adapt. These will include many in the developing world, low-lying islands and coastal regions, and the urban poor. The Framework Convention on Climate Change and the Kyoto Protocol represent the first steps taken by the international community to protect the Earth's climate from dangerous man-made interference. Currently, nations have agreed to reduce greenhouse gas emissions by an average of about 5 per cent from 1990 levels by the period 2008 to 2012. The UK, through its Climate Change Programme, has committed itself to a 12.5 per cent cut in greenhouse gas emissions. Additional commitments for further greenhouse gas emission reduction will need to be negotiated during the early part of the 21st century, if levels of greenhouse gas concentrations in the atmosphere are to be stabilised at reasonable levels. Existing and future targets can be achieved by embracing the concept of sustainable development - development today that does not compromise the development needs of future generations. In practical terms, this means using resources, particularly fossil-fuel-derived energy, more efficiently, re-using and recycling products where possible, and developing renewable forms of energy which are inexhaustible and do not pollute the atmosphere.

Global warming begins when sunlight reaches Earth. The clouds, atmospheric particles, reflective ground surfaces and ocean surface then reflected about 30 per cent of it back into space, while the remaining is absorbed by oceans, lands and air. This in turn heats the planet's surface and atmosphere, making life possible. As Earth warmed up, this solar energy is radiated by thermal radiation or infrared heat, travelling directly out to space, thus cooling the Earth. However, some of the outgoing radiation is re-absorbed by carbon dioxide, water vapour and other gases in the atmosphere and is radiated back to Earth's surface; these gases are known as greenhouse gases due to their heat-trapping capacity. This re-absorption process is naturally good; the Earth's average surface temperature would be very cold if not for the greenhouse gases.

GLOBAL WARMING: CAUSES AND RESULTING CLIMATE CHANGE

The most authoritative reports on the causes and consequences of climate change come from the IPCC, particularly its 1995 Second Assessment Report (SAR) and its 2001 Third Assessment Report (TAR). The latter report refined the findings of the first assessment, pointing out that climate change is likely to be worse and occur more rapidly than initially predicted. Here I summarize the IPCC's findings on global warming and the worldwide effects of climate change before pointing out some of the anticipated socio-economic impacts in East Asia.

According to the IPCC's TAR, there is now a collective picture, derived from an increasing body of observations, of a warming world and other changes in the Earth's climate system. The global average surface temperature increased during the twentieth century, with the 1990s and early 2000s the warmest on record. Snow and ice cover have decreased, global average sea level has risen, and the heat content of the oceans has increased.

Other aspects of climate have changed during the twentieth century, including changes in precipitation (e.g. increased heavy precipitation events) and cloud cover; fewer extreme low-temperature periods and more high- temperature periods; more frequent, persistent, and intense episodes of the El Nino ocean-warming event (and related adverse effects on weather in many areas); and an increase in areas experiencing drought and severe wet periods.

Some climate related events, such as tornadoes or tropical storms, do not appear to have changed based on IPCC data, although the evidence is conflicting. The TAR also finds that emissions of GHGs from human activities are altering the atmosphere in ways that are expected to affect climate.

Human activities have increased atmospheric concentrations of GHGs (e.g. CO_2, methane, nitrous oxide, halocarbons) and their warming potential. According to the report, "atmospheric concentration of carbon dioxide (CO_2) has increased 31 percent since 1750. The present CO_2 concentration has not been exceeded during the past 420,000 years and likely not during the past 20 million years. The current rate of increase is unprecedented during at least the past 20,000 years".

Three-quarters of human-induced emissions of CO_2 over the last two decades has come from the burning of fossil fuels (e.g. coal, oil, and natural gas), with most of the remainder the consequence of land-use changes, particularly deforestation. Natural causes of climate change have been relatively small. Furthermore, models for predicting future climate are increasingly accurate and precise. While uncertainties remain, understanding of climate processes and predicted effects has improved.

According to the TAR, new and stronger evidence points to human activities as the sources of observed global warming over the last fifty years, further strengthening the SAR's conclusion that the "balance of evidence suggests a discernible human influence on global climate". Warming over the last 100 years is unlikely to have been natural, with studies showing that global warming, particularly during the last 35-50 years, most likely resulted from human activities.

Thus, the TAR concludes: "In light of new evidence and taking into account the remaining uncertainties, most observed warming over the last fifty years is likely to have been due to the increase in GHG concentrations. Furthermore, it is very likely that the 20th century warming has contributed significantly to the observed sea level rise ... and widespread loss of land ice". Furthermore, the TAR determined that human activities will continue to shape the Earth's atmosphere throughout this century and into the future, and average global temperatures and sea levels are projected to rise. Emissions from burning fossil fuels will be the dominant source of atmospheric CO_2 during this century. These emissions and those of other GHGs would have to be reduced to "a very small fraction of current emissions" to stabilize climate.

Global average temperature is projected by the IPCC to increase by 1.4-5.8 degrees Celsius during this century (more than anticipated in the SAR). This warming will occur at a rate faster than that observed in the twentieth century, "very likely to be without precedent during at least the last 10,000 years". During this century, warming is expected to occur in most areas, but it should be particularly pronounced at northern high latitudes during winter.

Global mean sea level is expected to rise 0.09-0.88 metres in this century, with other very likely changes to include higher maximum temperatures and more hot days over most land areas, higher minimum temperatures and fewer cold days over most land areas, more intense precipitation events over many areas, increased summertime continental drying and drought over mid- latitude continental interiors, and more severe storms over some areas.

Ecological and Socio-economic Impacts of Climate Change

The ecological and socio-economic impacts of climate change are likely to be very significant and often painful. The TAR's findings on these impacts include the following: Regional changes in climate have already affected many physical and biological systems, with temperature increases being the most proximate cause.

Observed changes in regional climate have occurred in terrestrial, aquatic, and marine environments, and effects have included shrinking glaciers, thawing permafrost, reduced periods in which lakes and rivers are frozen, longer mid- and high-latitude growing seasons, shifts in animal and plant ranges to higher latitudes and altitudes, declines in populations of some animals and plants and reduced egg-laying in some birds, and insects populating new areas.

It appears that some social and economic systems have already been affected by increased floods and drought, but separating these ecological events from socio-economic factors is difficult.

The TAR shows that many human systems are sensitive to climate change, including water resources, agriculture, coastal zones and marine fisheries, settlements, energy, industry, financial services (e.g. insurance industries affected by increased claims), and human health.

Adverse impacts of climate change include reduced crop yields in most tropical and sub-tropical regions; decreased water availability in many water-scarce areas, especially the sub-tropics; more people exposed to increased mortality from heat stress and vector-borne diseases like cholera; widespread increase in flood risk from rising sea levels; and increasing demand for energy to cool areas affected by higher summer temperatures.

Some impacts may be positive, such as increased crop yields in some mid-latitude areas; potentially more timber if forests are managed appropriately (although increased pests could more than offset this); increased water availability for some water scarce areas; lower winter mortality in traditionally cold areas; and reduced winter demand for energy due to higher winter temperatures.

Many of the risks are unclear, and there is substantial potential for "large- scale and possibly irreversible impacts" from changing ocean currents, melting ice sheets, accelerated global warming due to atmospheric feedback effects, and so forth. In addition to efforts to mitigate climate change, the TAR argues that adaptation is a necessary strategy.

However, those people and societies with the least resources are most vulnerable because they are least able to adapt. Projected warming may result in a mixture of economic gains and losses for developed countries, but developing countries can expect mostly losses: "The projected distribution of economic impacts is such that it would increase the disparity in well-being between developed countries and developing countries," with the disparities increasing the greater the temperature increases. The upshot is that "More people are projected to be harmed than benefited by climate change," even if temperature increases are limited.

The TAR is not restricted to scientific and economic assessments. It argues that international justice and equity are important considerations when addressing climate change: "Inclusion of climatic risks in the design and implementation of national and international development initiatives can promote equity and development that is more sustainable and that reduces vulnerability to climate change".

Global Warming and Climate Change Impacts in East Asia

Clearly, the global effects of climate change are potentially major, and will likely lead to many adverse consequences, difficult choices, and expensive adaptation measures for much of the world's population. The countries of East Asia will not be immune to these changes, and in most cases will be among the worst affected due to their vulnerable geographies and economies.

Effects may not always be adverse, but even if they are not they will likely increase unpredictability and require adaptation. What are the expected impacts of climate change in East Asia? Several research reports have anticipated the effects of climate change for the region. Some of their findings are summarized here to convey the scale and nature of the potential changes.

According to a 1997 report from the IPCC on anticipated regional impacts of climate change, temperate Asia (including Japan, the Koreas, and most of China) has experienced an average annual temperature increase of more than 1 degree Celsius in the last century, mostly since the 1970s, with substantial warming expected in this century. Rainfall is expected to change in the area, with substantial declines expected in most of China (notably northern provinces). Permafrost in northeast China is expected to disappear (with release of methane, thus adding GHGs to the atmosphere) and glaciers will melt. Northern China is particularly vulnerable to expected changes in rainfall, exacerbating existing water shortages.

The area is likely to experience changing agricultural yields, with many crops likely to see reductions and a northward movement of crop zones and anticipated shortages of roundwood (partly due to increased demand). Delta coastlines in China "face severe problems" from sea-level rise, which will include salt water intrusion into aquifers. Japan will not be immune; already many parts of major coastal urban areas, with millions of residents, are below the mean high-water mark.

Providing protection for only some of these cities will cost tens of billions of dollars. Japan's beaches, which comprise about a quarter of its coastline, will be subject to erosion - and over half of existing beaches may disappear. Additionally, heat-related deaths throughout temperate Asia may increase sevenfold by the middle of this century.

The potential effects of climate change for tropical Asia (encompassing Southeast Asia) are also described in the IPCC's 1997 regional report. It points out that the region already suffers from increasing pollution, land degradation, and all manner of environmental problems resulting from rapid urbanization, industrialization, and economic development. Climate change will exacerbate these problems. In this area, mean temperatures have already gone up by 0.3-0.8 degrees Celsius over the last 100 years. Forest cover will change as a consequence, possibly increasing, and forest types may change. Changes in evaporation and rainfall are likely to have detrimental effects on freshwater wetlands. Coastal areas will be most greatly affected by sea-level rise and increased ocean temperatures (the latter possibly preventing coral reefs from keeping up with sea-level rise). Mangrove and tidal wetlands will have difficulty adapting due to bordering infrastructure and human activities. Greater erosion, coastal flooding, and salinization of fresh water sources are probable. Delta regions of Southeast Asian countries are particularly vulnerable, and throughout this area several million people could be displaced by sea-level rise.

The costs of responding to the impacts of rising seas, in the words of the IPCC, could be immense. Glaciers feeding the area's rivers will melt, and there may be yearly reductions - albeit between seasonal flooding - in the flow of snow-fed rivers, adversely affecting agriculture, hydropower generation, and urban water supplies.

Agriculture will probably suffer (despite CO_2 fertilization) from temperature and moisture changes and possibly from increased pests, affecting, for example, wheat, rice, and sorghum crops (although much uncertainty, confounding planning, will obtain). According to the IPCC report, poor rural populations depending on traditional forms of agriculture or living on marginal lands are especially vulnerable. Increased vector-borne diseases such as dengue, malaria, and schistosomiasis will adversely affect human health in this area.

A 1999 report on climate change impacts prepared by Britain's Climatic Research Unit summarized many potential impacts for some of the countries of East Asia. In China, temperature increases are predicted to be greatest over northern areas, with changes in precipitation and threats to biodiversity. In Indonesia, forest fires are predicted to increase and endangered species may be threatened.

In Japan, heat waves will increase in frequency and intensity, coastlines and coastal infrastructure will be harmed, and reefs will be stressed. In the Philippines, rainfall will increase during the wet season and decrease during the dry season, reefs will suffer from warming water, and potentially millions of people will be threatened by sea-level rise.

Von Hippel has summarized a few of the possible impacts of climate change in Northeast Asia: pressure on agricultural resources and accelerated desertification leading to cross-border migrations, particularly from China to Russia; adverse climatic effects on North Korea's food production, possibly increasing military pressure on South Korea and creating economic burdens for reunification; increased demand for air conditioning, leading to higher fuel consumption and hence more local and regional air pollution (and adding still further to GHG emissions); salinization of breeding grounds for fish from sea-level rise, leading to reduced fishery yields that could exacerbate conflicts over marine resources; additional oil pollution (from shipments of oil imports) that may strain relations among countries sharing marine resources and shipping lanes; and increased economic costs from natural disasters like catastrophic storms, straining emergency and disaster relief resources in the region.

The 2001 TAR assessment of vulnerability in Asia shows that the region is potentially more susceptible to climate change than are some other regions of the world. It concludes that the developing countries of Asia are highly vulnerable to climate change, and their adaptability is low. (Developed countries of the region (e.g. Japan) are of course less vulnerable because they are more able to adapt to climate change.)

Floods, forest fires, cyclones, droughts and other extreme events have increased in temperate and tropical Asia. The TAR anticipates that while agricultural productivity could increase in northern parts of Asia, food security would suffer in arid, tropical, and temperate Asia due to reduced agricultural and aquaculture productivity from warmer water, sea-level rise, floods, droughts, and cyclones.

Water availability may decrease in arid and semi-arid Asia and possibly increase in northern Asia, and increased incidence of vector-borne diseases and heat-stress will threaten human health. Temperate and tropical Asia should anticipate increased rainfall and floods, and sea-level rise and more intense storms could "displace tens of millions of people in low-lying coastal areas of temperate and tropical Asia". Some parts of Asia will see climate change effects on transport, increased demand for energy, and adverse impacts on tourism. Land-use and land-cover changes will threaten biodiversity, and sea-level rise will adversely impact coral reefs and mangrove areas that are important for fisheries.

What comes from these (and other) reports on the impacts of climate change in East Asia is that many of the effects will be felt most by - and be most painful for - the developing countries of the region. They are generally more vulnerable and least able to cope due to poverty and existing environmental problems and resource scarcities.

A very large number of people throughout East Asia live in low-lying coastal regions, and they are threatened by sea-level rise, land subsidence, inundation of fresh water aquifers by salt water, and more frequent and violent storms from climate change. Island countries such as Indonesia and the Philippines are especially vulnerable to climate change effects.

They can expect freshwater shortages and damage to coastal areas and adjacent infrastructure, with concomitant adverse effects on tourism. (Indeed, in extreme cases it may one day be necessary for some small-island states to abandon their territory altogether. Representatives from these countries have for some time argued that they are already feeling the effects of rising oceans.

Other poorer countries in the region are vulnerable. For example, the World Bank reported that Chinese research has estimated that a 1-metre rise in sea level would inundate 92,000 square kilometres of China's coast, displacing 67 million people (and more as population increases). According to one assessment, future climate change will reduce soil moisture in China, particularly in the north, and this will increase the demand for agricultural irrigation, which will in turn add to existing severe water shortages.

In short, "Possible impacts of climate change on Chinese agriculture could be highly disruptive ...". Already vulnerable, China may also see greater weather extremes, including droughts in the north and floods in the south, and heat stroke and death will increase, as may occurrences of malaria, dengue fever, and other diseases. An ever-growing body of research shows that climate change will (and probably has already) adversely affected human health, and this is particularly true of East Asia. Southeast Asia is especially vulnerable to anticipated increasing incidence of vector-born diseases. Hotter weather will increase heat-related mortality in the region, as indicated by historical studies from China showing a strong correlation between peak summer temperatures and death rates.

But even the developed countries and regions of East Asia are unlikely to avoid harm from climate change. For example, while Japan's coastlines are not as vulnerable as those of China, the Philippines, and other countries, it is reasonable to expect that it will suffer costly damage from sea-level rise, associated storm surges, and adverse weather, and it has direct interests in the health of surrounding seas and indirect interest in what happens throughout the region.

By way of example, recently reduced fish catches by Japanese fisherman have been attributed to changes in underwater currents triggered by global warming. And there will be adverse impacts for Japan's biodiversity, forests, agriculture, wetlands, and water systems, as well as for infrastructure and human health. According to the government, climate change effects have already become visible in Japan. For these and other reasons, Japan supports the climate change regime and the Kyoto Protocol- despite its greater ability to cope with climate change compared to its neighbours.

CONCENTRATION OF GREENHOUSE GASES IN THE ATMOSPHERE

The problem begins when the concentration of greenhouse gases in the atmosphere were artificially raised by humankind at an ever-increasing rate since the past 250 years. As of 2004, over 8 billion tons of carbon dioxide was pumped out per year; natural carbon sinks such as forests and the ocean absorbed some of this, while the rest accumulated in the atmosphere. Millions of pounds of methane are produced in landfills and agricultural decomposition of biomass and animal manure. Nitrous oxide is released into the atmosphere by nitrogen-based fertilizers and other soil management practices. Once released, these greenhouse gases stay in the atmosphere for decades or longer. According to the Intergovernmental Panel on Climate Change (IPCC), carbon dioxide and methane levels have increased by 35 and 148 per cent since the 1750 industrial revolution. Paleoclimate readings taken from ice cores and fossil records dating back to 650 000 years show that both gases are at their highest levels. Thermal radiation is obstructed further by the increased concentrations of greenhouse gases, resulting in what is known as enhanced global warming.

Recent observations of global warming have solidified the theory that it is indeed an enhanced greenhouse effect that is causing the world to warm. The planet has experienced the largest increase in surface temperature over the last century. Between 1906 and 2006, the Earth's average surface temperature rose between 0.6 to 0.9 degrees Celsius; the last 50 years saw the temperature increase rate almost doubling. Sea levels have shown a rise of about 0.17 metres during the twentieth century. The extent of Arctic sea ice has steadily shrunk by 2.7 per cent per decade since 1978, just as world's glaciers steadily receded.

As the world continues to consume ever more fossil fuel energy, greenhouse gas concentrations will continue to rise, and with them Earth's temperature. The IPCC estimates that based on plausible emission scenarios, average surface temperatures could increase between 2°C and 6°C by the end of the 21st century. Continued warming at current rates poses serious consequences. Low-lying coastal regions, with dense population, are especially vulnerable to climate shifts, with the poorer countries and small island nations having the hardest time adapting. It has been projected that by 2080, 13 to 88 million people around the world would lose their home to floods.

20th Century Climate Change

There is now considerable evidence that indicates a relationship between the man-made enhancement of the Earth's natural greenhouse effect through greenhouse gas pollution of the atmosphere and the global warming that has been observed.

Measurements of surface temperature recorded around the world during the last 150 years indicate that global temperatures are now higher than in any decade over this period. During the 20th century a global average surface temperature increase of about 0.6°C years has taken place, although the warming trend has not been smooth and has taken place rather differently between the Northern and Southern Hemispheres. With this in mind, it is important to recognise that global average surface temperature, as a measure of the global climate, represents an over-simplification. Winter temperatures and night-time minimums for example, may have risen more than summer temperatures and daytime maximums.

With higher global temperatures one would expect an increase in rainfall and other forms of precipitation, because of the greater amount of moisture available within the atmosphere. Striking changes in precipitation have occurred on regional scales, most notably the drought in the African Sahel between the 1960s and 1980s. Nevertheless, the accuracy of many precipitation records should be treated with caution. Precipitation is more difficult to monitor than temperature due to its greater geographical variability.

Variations in land- and sea-ice coverage and the melting or growth of glaciers occur in response to changes in temperature, sunshine, precipitation, and for sea-ice changes in wind. Since 1966 Northern Hemisphere snow cover maps have been produced by the United States using satellite imagery. Consistent with the surface and tropospheric temperature measurements is the decrease (by about 10 per cent) in snow cover and extent since the late 1960s. There has been a widespread retreat of mountain glaciers in non-polar regions during the 20th century. Variations in sea-ice extent have also been reported, with spring and summer sea-ice extent in the Northern Hemisphere decreasing by between 10 an 15 per cent since the 1950s. Considerable interest is now focusing on Antarctica, where regional warming, as predicted by climate models, has been more rapid than global warming as a whole. In recent years the summertime disintegration of the Larsen Sea Ice Shelf adjacent the Antarctic continent has been occurring on an unprecedented scale. In view of the rapidity at which it is taking place, such an event has been viewed as a possible signal of global warming.

Increased global cloudiness, as for increased global evaporation and precipitation, would be an expected consequence of higher global temperatures. It is likely that there has been a 2 per cent increase in cloud cover over mid - to high latitude land areas during the 20th century. In most areas the trends relate well to the observed decrease in daily temperature range (since cloudier nights tend to be warmer and cloudier days cooler).

21st Century Climate Change

Prediction of climate change over the next 100 to 150 years is based solely on climate model simulations run on computers. The vast majority of modelling has concentrated on the effects of continued man-made pollution of the atmosphere by greenhouse gases, and to a lesser extent, atmospheric aerosols. The main concern at present is to determine how much the Earth will warm in the near future.

Significant results from some of the best climate models available indicate that a global average warming of 0.3°C per decade can be expected to occur during the 21st century, assuming that mankind fails to control current emissions of greenhouse gases, although it could be as high as 0.6°C. In addition regional variations in the patterns of temperature and precipitation change will occur, with greater warming likely in the polar regions. Currently, models suggest that if the atmospheric concentration of carbon dioxide, the main greenhouse gas, is doubled from pre-industrial levels, the Earth will warm by between 1.5 and 4.5°C sometime over the next 200 years or so. The large margin of error in future prediction of temperature emphasises that modelling the climate is inherently a difficult business. Part of the problem stems from trying to guess what climate feedbacks might occur that may enhance the initial warming due to an enhanced greenhouse effect. Melting ice in the polar regions for example could accelerate warming because exposed ground absorbs more energy from the Sun than snow and ice, which reflect about 80 to 90 per cent.

Whilst uncertainties concerning the actual response of the global climate to man-made greenhouse gas emissions exist, most scientists agree that the global warming trend of the 20th century will continue into the 21st century. The projected rate of warming is faster than at any time during recent Earth history. If nations fail to respond, the world may experience numerous adverse impacts as a result of global warming in the decades ahead.

ATMOSPHERIC AEROSOLS

Atmospheric aerosols are very fine particles suspended in air. They are formed by the dispersal of material at the Earth's surface (primary aerosols), or by reaction of gases in the atmosphere (secondary aerosols). They include sulphates and nitrates from the oxidation respectively of sulphur dioxide and nitric oxide during the burning of fossil fuels, organic materials from the oxidation of volatile organic compounds (VOCs), soot from fires, and mineral dust from wind-blown processes. Natural aerosols, which also include sea salt and volcanic dust, are probably 4 to 5 times larger than man-made ones on a global scale, but regional variations in man-made pollution may change this ratio significantly in certain areas, particularly in the industrialised Northern Hemisphere. Although making up only 1 part in a billion of the mass of the atmosphere, they have the potential to significantly influence the amount of sunlight reaching the Earth's surface, and therefore climate. Removal of most aerosols is mainly achieved by rainfall (wet deposition) and by direct uptake at the surface (dry deposition). Explosive volcanic eruptions however, can inject large quantities of dust and gaseous material, such as sulphur dioxide, high into the atmosphere (the stratosphere). Here, sulphur dioxide is rapidly converted into sulphuric acid aerosols. Whereas pollution of the lower atmosphere is removed within days by the effects of rainfall and gravity, stratospheric pollution may remain there for several years, gradually spreading to cover much of the globe.

Like greenhouse gases, aerosols influence the climate. Atmospheric aerosols influence the transfer of energy in the atmosphere in two ways: directly through the scattering of sunlight; and indirectly through modifying the optical properties and lifetimes of clouds. The scattering of sunlight by aerosols is clearly demonstrated in the aftermath of a major volcanic eruption, when exceptionally colourful sunsets may be witnessed. The volcanic pollution results in a substantial reduction in the direct solar beam, largely through scattering by the highly reflective sulphuric acid aerosols. Overall, there is a net reduction of 5 to 10 per cent in energy received at the Earth's surface. An individual eruption may cause a global cooling of up to 0.3°C, with the effects lasting 1 to 2 years.

Estimation of the impact aerosols have on longer-term global climate change however, is more complex and hence more uncertain than that due to the well-mixed greenhouse gases. This is largely because the geographical distribution of aerosols is highly variable and strongly related to their sources. The best estimates of global cooling attributable to man-made aerosols are based on computer models. These show that the global cooling effect of man- made aerosols could offset the warming effect of increased greenhouse gas concentrations by as much as 30 per cent. The variable distribution of aerosols however, makes calculation of a global average difficult. Nevertheless, it is likely that aerosols may slow the rate of projected global warming during the 21st century.

Agriculture

Climate is the most significant factor in determining plant growth and productivity. Without intervention to reduce emissions of greenhouse gases, global average surface temperature is projected to increase by about 0.2°C per decade during the 21st century. This rapid change in climate will have major implications for agriculture around the world. Crop growth is often limited by temperature. Temperatures during the 21st century are expected to increase more in the higher latitudes where shifts in vegetation will be greater. In Britain an increase in temperature of 1.5°C by 2050 is the equivalent of a decrease in altitude of approximately 200m.

This is the same as a shift southward in latitude of 200-300 km. Such an increase in temperature would allow widespread maize cultivation across southern England to take place. In other regions however, a rise in temperature may not be so beneficial. Small increases in temperature would extend the range of temperature-limited pests. The European Corn Borer for example, a major pest of grain maize, may shift between 165 and 500 km northwards with a rise of 1°C.

Moisture and water availability will be affected by a temperature increase, regardless of any change in rainfall. Higher temperatures increase the evaporation rate, thus reducing the level of moisture available for plant growth, although other climatic elements are involved. A warming of 1°C, with no change in precipitation, may decrease yields of wheat and maize in the core cropping regions such as the US by about 5 per cent. A very large decrease in moisture availability in the dryer regions of the world would be of great concern to the subsistence farmers that farm these lands. Reduced moisture availability would only exacerbate the existing problems of infertile soils, soil erosion and poor crop yields. In the extreme case, a reduction in moisture could lead to desertification.

Sea levels have been projected to rise by anywhere up to a metre by 2100, although considerable uncertainty is attached to this. The greatest threat to low-lying agricultural regions from sea level rise is that of inundation and flooding. Southeast Asia would have the greatest threat of inundation because of the deltaic nature of the land. Furthermore, the pollution of surface and groundwater with salty seawater is another potential problem facing farmers situated in low-lying regions. The costs of agricultural production would increase, resulting in higher food prices for the consumer. Although climate changes may have some detrimental impacts on agricultural production around the world, the increase in atmospheric carbon dioxide concentrations could be beneficial. Plants grow as a result of photosynthesis - the mechanism whereby the plant converts carbon dioxide from the atmosphere into food. With higher levels of carbon dioxide stimulating the rate of photosynthesis, the growth rate and productivity of plants could be expected to increase. This would be beneficial for global food stocks.

Most crops grown in cool, temperate regions respond positively to an increased concentration of carbon dioxide, including some of the current major food staples such as wheat, rice and soybean. Some studies have shown that growth rate in these crops may increase up to 50 per cent if carbon dioxide in the atmosphere is doubled. Crops grown in the tropical regions of the world, including sorghum, maize, sugar cane and millet, which together account for about one fifth of the world's food production, do not respond as well to increases in carbon dioxide. In order to maintain agricultural output to meet the demand for a growing world population, farmers will have to adjust and adapt as and when necessary to the possible changes imposed by changing climate. Higher temperatures would increase the demand for irrigation of agricultural land.

Unfortunately, in many arid and semi-arid regions of the world the demand for water already exceeds supply. Increased spread of pests and disease may also place additional demands on the need for fertilizers, pesticides and herbicides which are costly. The ability to adapt to the effects of climate change will vary greatly between countries and regions. Economic and technological constraints will limit the rate of adaptability, with poorer economies lagging behind. Consequently, without intervention the effects of climate change in the 21st century look set to further widen the gulf between developed and developing nations.

EVIDENCE FOR GLOBAL WARMING

Degradation of Earth's Atmosphere; Temperature Rise; Glacial Melting and Sealevel Rise; Ozone Holes; Vegetation Response "Global change". "Greenhouse effect". "Global warming". The media are full of statements, concerns, guesses, and speculation about these phenomena, as scientists and policy-makers around the world struggle to address recent scientific observations that indicate human activities impact our environment. And yet, each of these is a "natural" phenomenon, as are many others. Hurricanes, droughts, and monsoons all occur without any control by humans, to initiate, forestall or moderate them. Most readers of this Tutorial already know that global warming is litreally a "hot" topic. For those still with relative unfamiliarity, we recommend this Wikpedia review: effects of global warming.

This page considers the nature of and evidence for global warming. There seems almost no doubt that the Earth's climate has been steadily warming overall in the last hundred years: global warming is a fact - a reality. The nagging question - still unanswered to everyone's satisfaction - is: How much of the quantified increase of about 1° C in that time interval is just due to a natural trend probably related to glacial cycles (which themselves are temperature-dependent) and how much is additional warming accelerated by human activity as the industrial revolution reaches rapidly increasing levels. The present page provides an overview. Since it was written, the writer (NMS) has encountered new lines of evidence - pro and con.

Most people who follow the news accounts of global warming have come to realise that the primary culprit is human contributions of heat-affecting gases into the atmosphere as these are released by everyday processes such as burning coal to produce electricity and consuming gasoline to run automobiles. The most blamed gas is carbon dioxide (CO_2). The United States is often cited as the principal polluting nation, although now China and India - each with more than a billion people, many of whom are now driving cars - are becoming major players. In the U.S., cities and surrounding metropolitan areas are the chief sources of CO_2, as evident in this map made from ground and aerial sounding monitors but also containing some input from satellite measurements:

To fully understand global warming, one must first understand the operations of the Earth System, since natural sources of temperature-affecting gases play an important role. We can learn about our planet's interacting physical systems by observing the results of such natural phenomena, and use our knowledge to explore human-induced changes. Thus, one factor that may in itself induce and account for some of the regional and global temperature changes is volcanism. Remote sensing is effective at monitoring such events. Consider, for example, the eruption of a volcano, such as Mount Pinatubo in the Philippine Islands in 1991, that happened without human intervention. This volcano had been dormant for more than 600 years.

When a volcano erupts it spews millions of tons of ash, debris, and gases into the atmosphere, not to mention the lava flows from some volcanoes. Because of the presence of instruments - on the ground, at the ocean's surface, and in space - meteorologists/environmentalists observed a cloud of sulfur dioxide (SO2), emitted by Pinatubo, make its way westward, extending well past India within twelve days of the original eruption.

Monitoring has been done by the UARS (Upper Atmosphere Research Satellite). By three months, that cloud had completely encircled the Earth, as shown from space (below), and inside of a year SO2 particles in the atmosphere were providing gloriously-coloured sunsets all over the globe and lowering global temperatures, as well. Clearly, an erupting volcano impacts more people and places than just within its immediate vicinity.

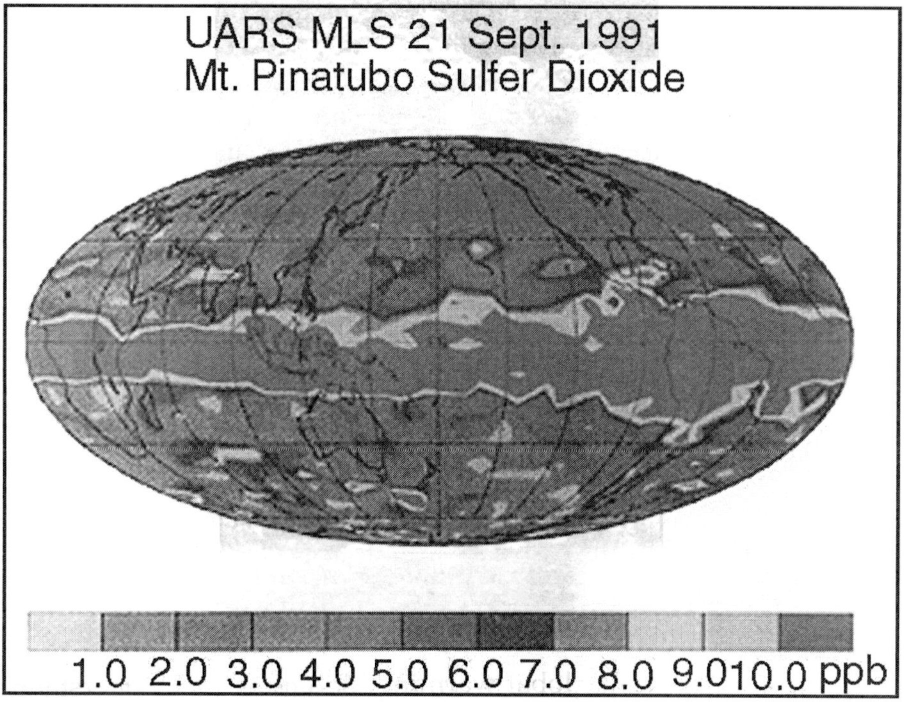
UARS MLS 21 Sept. 1991
Mt. Pinatubo Sulfer Dioxide
1.0 2.0 3.0 4.0 5.0 6.0 7.0 8.0 9.0 10.0 ppb

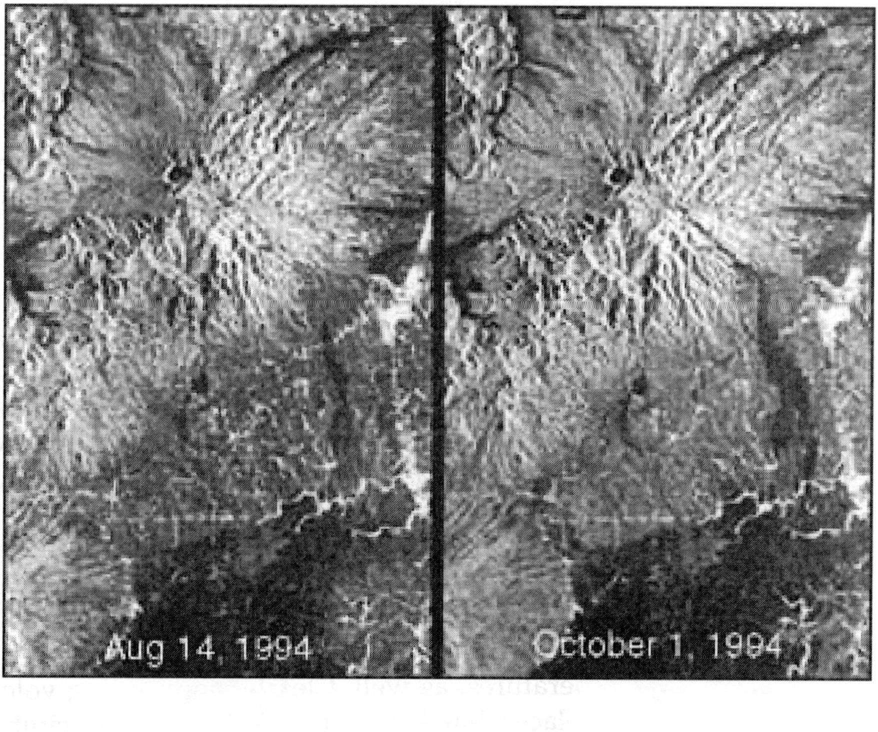
Aug 14, 1994
October 1, 1994

Volcanoes are a main source of SO_2 rises in atmospheric chemistry. The TOMS instrument has been making measurments of SO_2 in the air for more than 30 years now. Another fascinating example of a natural phenomenon we know as El Niño, because it occurs with some regularity (although not complete predictability) around Christmas time. El Niño refers to the baby Jesus, whose birthday we celebrate at the end of December. When an El Niño occurs, a pool of warm water from the western Pacific Ocean moves eastward to the western coast of South America. In the process, weather patterns around the world changes often to the detriment of human populations–as do South American fish populations. In non-El Niño years, fish are abundant in this region, because of the cold, nutrient-filled waters. When an El Niño occurs, that cold water flows deep into the Pacific, and fish populations decline dramatically, with concomitant effects on humans whose livelihood depends on those fish.

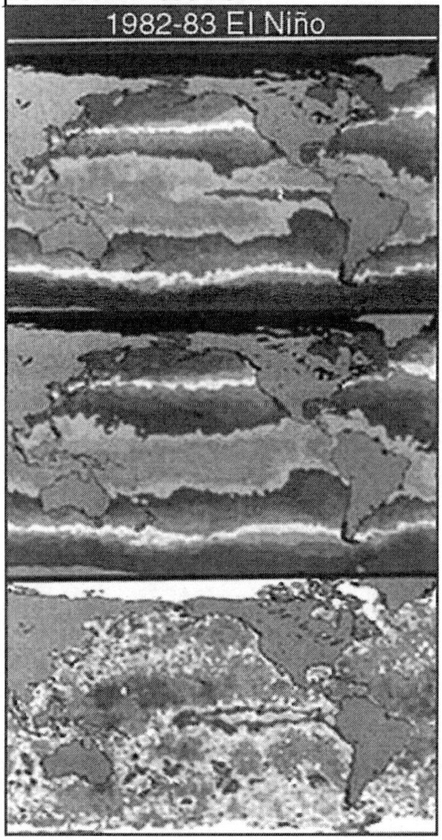

Other such phenomena abound. Some are readily observable by space sensors, particularly meteorological and oceanographic measurements. As examples, consider these near-global plots. The top from comes from the Seasat Radar Scatterometre and shows prevailing wind patterns over the oceans at supercontinental scales. The bottom one depicts the mean day, night, and day- nite temperatures of the Earth's land and sea surfaces, averaged for January of 1979, from Nimbus 6's High-resolution Infra-red Sensor (HIRS) 3.4 and 4.0 μm channels, integrated with MSU microwave and infrared data.

Suffice to say that even without human contributions Earth is a dynamic system, one that changes routinely and often drastically. Clearly, global mean annual temperatures are rising, and we continue to monitor this condition with space observations to help settle the question: how much is just a natural trend (e.g., inevitable interglacial warming) and how much is due to man's activities?

Calculations show that the burning of fossil fuels (mainly coal, petroleum derivatives, and natural gas) add about 6 billion metric tons of carbon (as the element) to the air annually; each year also, deforestation permits an extra 1-2 billion metric tons of carbon to reach the atmosphere. As an indication of how much worldwide temperatures have risen in the last few years, this next map of the globe shows the geographic distribution of temperature anomalies, with measurements from various sources as compiled by NASA GISS (Goddard Institute for Space Science), for the year 2006 - the fifth warmest on record.

The plot of temperature increase since 1880 is shown but is hard to read. So, beneath it is a recent version that clearly presents details: GISS is continually refining its models. Daily inputs improve the credibility of the predictions. Now that Aqua is operational, its AMSR instrument can provide global data covering very short time spans. This map is of worldwide surface temperatures on August 27, 2003:

Using temperature data, GISS has now published a map of sea surface heat content anomalies (measured to a water depth of 70 m) integrated over a ten year period from 1993 to 2003. From typical data they have derived a model distribution of temperature variations. Since the upper ocean waters serve as efficient reservoirs for heat storage, the patterns of distribution of excess, normal, and deficient heat (in Watts-year/m2) thus determined are proving helpful in weather and climate forecasting since the temperatures have strong effects on atmospheric heating.

The trend of temperature changes on annual and 5 year-running averages proves that in the second half of the 20th Century the global atmosphere has been slowly warming - cumulatively totaling about 0.5° C, as shown here:

Under the 2085 conditions, rainfall in this region will diminish and vegetation will become much like that around Phoenix, AZ today. Available water will become scarce, so that desalination of the Atlantic ocean would have to provide much that is needed.

That this warming trend is already happening elsewhere in North America is convincingly demonstrated by temperature data for the state of California. In the illustration, temperatures (in Fahrenheit) across the state and in Los Angeles have risen more than 2.5° F in places in the last four decades and more generally almost 4° F statewide in the last 120 years:

So, is there any sign this is happening everywhere? Much debate has started in the middle of 2006 because of very intense heat waves during July- August in both the U.S. and Europe (the latter seldom experiences the extremes that occurred then) as shown in these plots obtain from metsats: These excess temperatures point up the problem that both advocates and disbelievers of global warming face in interpreting the abnormalities. Thus: Does this period of high temperatures (St. Louis, MO had almost a week of temperatures at or above 100° F) reflect the onset of the predicted hotter summers or is this just another of the drastic warm spells that have happened infrequently in the past? The "jury is still out", although the unusual European situation seems to favour the warming hypothesis.

In addition to the observed phenomena described above, scientists have accumulated other relevant environmental recordings. For example, levels of several trace gases in our atmosphere have been rising and continue to rise. This change in atmospheric gas composition may be the most serious threat to the global well-being of Earth's environment.

Information about this has been nicely summarized in an October, 1997 pamphlet entitled *Climate Change: State of Knowledge*, prepared and distributed by the President's Office of Science and Technology (OST), from which information and selected illustrations on this page have been extracted.One of these gases, carbon dioxide (CO_2) has been increasing at an accelerating rate since the middle of the last century, after many, many years of essentially stable levels. Why is this? A simple explanation is that the Industrial Revolution began at about the time these increases started. With that social upheaval came the use of biomass and coal for fuel to support these new industries. Burning such material generates CO_2.

Other trace gases have been rising, as well. Carbon Monoxide (CO) is particularly deletorious. Methane (CH_4), from rice paddy production and enteric fermentation, is increasing, as are chlorofluorocarbons (CFCs) that have been used for many years as a refrigerant and to produce foam. Methane is a much more potent greenhouse gas that CO2.This pie chart shows the relative contributions of the different gases involved in atmospheric contamination:

As you will learn shortly, CO_2 is generally considered to be the prime culprit in any supposed global warming. For the next few paragraphs, we will divert from this theme to review some basic facts that provide a background for discussing CO2's role in moderating the atmosphere:

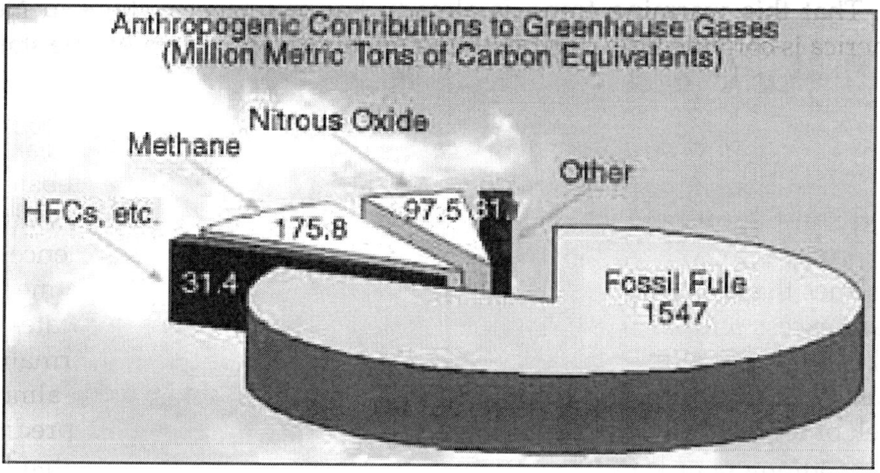

As we saw subsection that presented a "short course" in meteorology, the Earth's atmosphere obtains almost all its thermal energy from the Sun. This diagram is based on the general blackbody radiation curve.It shows the peak wavelength for radiation input from the Sun and the peak wavelength of the Earth itself as a thermal body of average temperature of 288 Kelvin. (The two curves have the same height in the plots; in fact the height for Earth is greatly subdued, indicating its energy output is much lower.) The amount of carbon dioxide added or subtracted through natural processes (controlled mainly by vegetation) is seasonal but the overall level as annualised is fairly constant.

In almost all discussions of global warming caused by CO_2, the plots used to demonstrate this effect show temperature versus time (years).

This results in two kinds of plots:

1. One that shows variations (ups and downs) over long periods; and
2. One that shows a steady increase in recent years.

This is pertinent information but a third plot would be equally informative: one that simply shows how temperature increases with concentration (the amount of CO_2 in a fixed volume). This is a straightforward type of measurement in Physics.

The irrefutable message from this plot: experimentally, it is clearly demonstrated that the addition of increasing amounts of CO_2 to an otherwise constant atmosphere causes a systematic rise in gas temperature. The ranges shown in the above plot are pertinent to levels that might be attained in the future if carbon dioxide emissions are not curbed.

Now, back to a survey of information relevant to global warming. Later we will familiarize you with dedicated satellites that gather data on atmospheric gases and other contributors to global warming. This bears on a main thesis in this Tutorial, that satellites are proving to be a powerful, almost unique, tool for monitoring the whole Earth at the global scale.

These new data sets will be invaluable. Scientists need to establish a long-term data base for global warming, which includes temperatures from previous centuries. One way to estimate past temperatures is from CO_2 measurements of air trapped in glacial ice that has been accumulating for thousands of years.

Mankind can probably live with these changes as we adjust and find ways to accommodate them. But the Gore book describes some climate-influenced events that will be devastating and considers certain possible effects that may be catastrophic. The many varied observations described on this page seem to favour the global warming hypothesis. But we must reiterate one statement made above: This warming may just be (largely) a natural effect of an interglacial temperature process (but nevertheless there clearly seems to be a component of the observed temperature increase that correlates well with humans' adverse contributions). If so, not to worry excessively (but some protective response is in order); if not, then the Gore conclusion - that very serious damage to global environments is about to occur - should be a warning that seems to cry out for a quick and drastic response by the planet's polluting nations.

In early 2007, a panel of more than a thousand scientists released a pair of reports summarizing what Science can state with high reliability about global warming.

To paraphrase some of their conclusions:

- Over the past few million years, natural climate changes, driven mainly in response to factors that cause widespread glaciation and alternate interglacial periods, are marked by intervals of slowly rising temperatures (interglacials) and then falling temperatures (continental glaciation). Rates of change are usually less than 1 degree per thousand years, and the total range is probably less than 20° F.
- During interglacials, some animal or plant species adapted to colder weather have died off (the same may happen during warm weather; new species can develop during either cold or warm climes).
- During interglacials, sealevel tends to rise and vegetation distribution varies; many biomes migrate poleward; productivity of edible plants may increase or decrease.
- It is likely that mankind has blossomed since the end of the last glaciation and has become both numerous and dispersed in the last ten thousand years. Now humans are in such abundance as to begin to affect (impact) natural conditions that control the

weather, moderate the environment, and otherwise maintain healthy living conditions.

- In all probability, some of the temperature rise whose rate is increasing is just a consequence of the natural interglacial warming. But a proportion of that rise (amount still guesswork) is almost certainly the influence of detrimental atmospheric gases that are being added from fossil fuel burning and other sources. The rise may be abnormal - it is the sum of both natural and manmade contributions.
- Since the human factor was not present in earlier interglacial warming trends, its current role in causing global warming cannot be evaluated in terms of outcomes. Many scientists believe that the rate and extent of temperature rise can be excessive enough to cause global equilibrium (adjustments of life to the changes) to be upset, perhaps to a stage that will lead to catastrophe. Work must be done to predict what might happen.
- Because of the uncertainties, the scientists strongly urge that mankind plays it safe and reduces or eliminates the known causes of atmospheric warming.

DETECTION OF ANTHROPOGENIC GLOBAL WARMING

The anthropogenic increase in atmospheric concentrations of greenhouse gases, and the associated increase in greenhouse radiative forcing. examined the observed climate changes, principally in surface temperature, that have taken place during the last 100 to 140 years. In this, the evidence for a causal link between the anthropogenic increase in greenhouse radiative forcing and the observed global warming will be reviewed.

The word "detection" by climate scientists has been used to refer to the identification of the significant change in climate during the twentieth century and its association with the anthropogenically enhanced greenhouse effect. Until 1995 most reviews had concluded that the enhanced greenhouse effect has not yet been detected unequivocally in the observational record. However, they have also noted that the global-mean temperature change over the past 100 years is consistent with the greenhouse hypothesis. Since then, the most recent scientific report from the Intergovernmental Panel on Climate Change has proposed that the balance of [modelling] evidence suggests a discernible human influence on the global climate does exist.

Traditionally, the difficulty in greenhouse detection has arisen because there are numerous other causes of climatic variability, and some of these [forcing mechanisms] may be operating on time scales (101 to 102 years) comparable to that of the anthropogenic greenhouse forcing. In addition, Wigley and Raper have shown that the inherent variability (random fluctuations) of the climate system can produce warming or cooling trends of up to 0.3°C per century.

The detection problem can be conveniently described in terms of "signal" and "noise". The signal here is the time-dependent climatic response to greenhouse forcing, whilst the noise is any non-greenhouse climate variation, either periodic, quasi-periodic or random.

Attribution and the Fingerprint Method

In the last part, it was recognised that although a good case for linking the enhanced greenhouse effect to global warming could be proposed, a high degree of confidence could not be attached to such a cause-and-effect relationship from studies of a single variable (surface temperature). Linking cause and effect is referred to as attribution. Confidence in the attribution is increased as model predictions of changes in various components of the climate system are borne out by the observed data in more and more detail. This method is known as the fingerprint approach; namely, identification of an observed signal that has a structure unique to the predicted enhanced greenhouse effect. It was noted that the use of global-mean surface temperature as a fingerprint variable does not permit attribution.

The fingerprint method is essentially a form of model validation, where the perturbation experiment that is being used to test the models is the currently uncontrolled emissions of greenhouse gases into the atmosphere. In addition to global-mean surface temperature, there are a number of other fingerprint variables that can be used. In choosing them, three issues must be considered:

- The signal-to-noise ratio - should be maximised;
- Uncertainties in both the predicted signal and the noise - these should be minimised, and;
- The availability of suitable observational data.

Latitudinal Surface Temperatures

Most simulations suggest that the warming north of 50°N in the winter half of the year should be enhanced due to ice-albedo feedback mechanisms. Over the last 100 years, high northern latitudes have warmed slightly more than the global mean during winter, but since the 1920s there is little noticeable trend. More importantly, however, high-latitude warming does not necessarily represent a unique fingerprint of greenhouse warming. Ice-albedo feedbacks could be equally invoked for other causes of climate change.

Tropospheric Warming and Stratospheric Cooling

All equilibrium model simulations show a warming in the mid-troposphere and cooling in the stratosphere. It has been suggested that this contrast in trends between the troposphere and stratosphere might prove a useful detection fingerprint. In addition, high signal-to-noise ratios have been obtained for free tropospheric temperatures. However, stratospheric cooling may not solely be attributed to greenhouse forcing, and can arise due to volcanic pollution injections and ozone depletion. Model simulations are also inconsistent with recent observations, in particular at the level at which warming reverses to cooling. Perhaps most significantly of all, though, is the lack of reliable instrumental data. The instrumental record for upper atmosphere temperatures extends back only to the late 1950s with the use of radiosonde, whilst satellite-based data is even younger.

Global-Mean Precipitation Increase

GCM models suggest an increase in global-mean precipitation, as one might expect from the associated increase in atmospheric temperature. However, because the spatial variability of rainfall is much greater than for surface temperature, regional and local details of changes are highly uncertain. Instrumental data from which long-term changes in precipitation can be determined are only available over land areas data suffers from the problems of incomplete coverage and inhomogeneity.

With the likelihood that the precipitation signal-to-noise ratio is low, any meaningful comparison between observation and model, and therefore attribution, is precluded. In addition, since global precipitation is partially dependent upon global temperature, rises in global precipitation may be expected as a consequence of other causes of climate change.

Sea Level Rise

Increasing greenhouse gases are expected to cause a rise in the global-mean sea level, due partly to oceanic thermal expansion and partly to the melting of land-based ice masses. However, as for global precipitation, such a variable is not wholly independent of global temperature. Thus, whilst both thermal expansion and glacial melting are consistent with global warming, neither provides any independent information about the cause of the warming.

Multivariate Fingerprints

Although most univariate detection methods of global climate change due to an enhanced greenhouse effect have their limitations, it is likely that multivariate fingerprint methods, which involve the simultaneous use of several time series, will facilitate attribution. In its most general form one might consider the time evolution of a 3-D spatial field, comparing model results with observations. This may be achieved by comparing changes in mean values and variances, or correlating spatial patterns between simulation and observation.

When will Attribution Occur

The fact that the cause (the enhanced greenhouse effect) and effect (global warming) have not unambiguously been linked leads to the question: when is this [attribution] likely to occur? Detection is not a simple yes/no issue. Rather, it involves the gradual accumulation of observational evidence in support of model predictions.

The scientist's task is to reduce the uncertainties associated with the understanding of the climate system. Thus, while the ultimate objective of climate research is detection, it is really the climate sensitivity one is interested in. A better understanding of this key parametre will not only allow one to argue with increased confidence that anthropogenic global warming is (or is not) occurring, but it will support prediction of future climate change, both anthropogenic and natural.

Future Climate Change

The issue of climate change detection introduces the wider concern of future climate change. Although predictions of climate change thousand, millions and even hundreds of millions of years from now could be made based on our knowledge of longer term climate changes, I will be concerned only with the current decadal to century time scale. Clearly, prediction of climate change over the next 100 to 150 years is based solely on model simulations.

Understandably, the vast majority of modelling has concentrated on the effects of continued anthropogenic pollution of the atmosphere by greenhouse gases, and to a lesser extent, atmospheric aerosols. The main concern, at present, is to determine how much the Earth will warm in the near future.

GCM Climate Simulations

During the last decade or so, a number of complex GCMs have attempted to simulate future anthropogenic climate change. Earlier models studied equilibrium climate change associated between two, presumably stable, states of climate. The results of these are discussed in the first IPCC report. Associated with a doubling of pre-industrial atmospheric CO_2, the following conclusions have been made:

- A global average warming at or near the Earth's surface of between 1.5 and 4.5°C, with a "best guess" of 2.5°C, will occur;
- The stratosphere will experience a significant cooling;
- Surface warming will be greater at high latitudes in winter, but less during the summer;

- Global precipitation will increase by 3 to 15 per cent;
- Year-round increases in precipitation in high-latitude regions are expected, whilst some tropical areas may experience small decreases.

More recent time-dependent GCMs which couple the atmospheric and oceanic components of the climate system together, provide more reliable estimates of greenhouse-gas-induced climate change. For steadily increasing greenhouse forcing, the global rise in temperature is typically less than the equilibrium rise corresponding to an instantaneous forcing. Significant results indicate:

- A global average warming of 0.3°C per decade, assuming non-interventionist greenhouse gas emission scenarios;
- A natural variability of about 0.3°C in global surface air temperature on decadal time scales;
- Regional patterns of temperature and precipitation change similar to equilibrium experiments, although warming is reduced in high latitude oceans where deep water is formed.

The ability to model the time-dependent nature of the climate system more adequately has allowed scientists to investigate the damping effects of the oceans on climate change.

Because the response time of the oceans, in particular the deep ocean, is much longer than for the free atmosphere, they have a regulating or delaying effect on the warming associated with enhanced greenhouse forcing. In addition, the transient GCMs have allowed increased attention to focus on the critical role of feedback processes in determining the climate's response to forcing perturbations.

Greenhouse Feedbacks

GCMs have estimated the CO_2 doubling temperature change in the absence of feedback processes to be approximately 1.2°C. The existence of feedback loops within the climate system results in a climate sensitivity of 1.5°C to 4.5°C, and makes an otherwise linear association between radiative forcing and global temperature distinctly non-linear. Three of the most important direct climatic feedbacks to greenhouse forcing include the water vapour feedback, the cloud feedback and the ice-albedo feedback.

Water Vapour Feedback

The importance of water vapour feedback in climate change has long been recognised. The concentration of water vapour in the atmosphere increases rapidly with rising temperature (about 6 per cent/°C); this is the basis for the strong positive water vapour feedback seen in current climate models (whereby increases in temperature produce increases in atmospheric water vapour which in turn enhance the greenhouse forcing leading to further warming). All current GCMs simulate a strong positive water vapour feedback and are in agreement with observational data.

Cloud Feedback

Cloud feedback is the term used to encompass effects of changes in cloud and their associated radiative properties on a change of climate, and has been identified as a major source of uncertainty in climate models. This feedback mechanism incorporates both changes in cloud distribution (both horizontal and vertical) and changes in cloud radiative properties (cloud optical depth and cloud droplet distribution); these are not mutually independent. Although clouds contribute to the greenhouse warming of the climate system by absorbing more outgoing infrared radiation (positive feedback), they also produce a cooling through the reflection and reduction in absorption of solar radiation (negative feedback).

It is generally assumed that low clouds become more reflective as temperatures increase, thereby introducing a negative feedback, whilst the feedback from high clouds depends upon their height and coverage and could be of either sign.

Greenhouse Gas Feedbacks

In addition to these climatic feedbacks, there are other non-climatic feedbacks which may enhance or diminish the increase in atmospheric greenhouse gas accumulation. Potentially, there are numerous feedback processes which may act on the carbon cycle, thus affecting the transfer of carbon dioxide between the various components. Carbon storage in the oceans will be affected by both changes in water temperature and ocean circulation resulting from global warming. Warmer water stores less CO_2, and future atmospheric CO_2 increase may be amplified by something like 5 to 10 per cent by this effect. A slow down in vertical overturning within the oceans will reduce the uptake of atmospheric CO_2. In addition, reduced nutrient recycling would limit primary biological productivity in the surface oceans, thereby increasing further the surface water partial pressure of CO_2, and consequently atmospheric CO_2.

Carbon cycle feedbacks involving terrestrial carbon storage may include carbon dioxide fertilisation of plant photosynthesis (a negative feedback); eutrophication involving nitrogen and phosphate nutrient fertilisation (a negative feedback); temperature feedbacks on the length of growing season and photosynthesis/respiration rates (both positive and negative feedbacks) and changes in the geographical distribution of vegetation. The methane gas cycle may also experience feedback effects. Recent attention has been focused on northern wetlands and permafrost, where increases in temperature and soil moisture would result in significant increases in methane release into the atmosphere.

Despite the increased ability of transient GCMs to model climate feedback processes, there is no compelling evidence to warrant changing the equilibrium sensitivity to doubled CO_2 from the range of 1.5 to 4.5°C as estimated by IPCC, 1990a.

GREENHOUSE MODELLING VERSUS OBSERVATION

Global-mean temperature has increased by around 0.3 to 0.6°C over the past 100 years. At the same time, greenhouse gas concentrations, and atmospheric aerosol loadings have increased substantially. To assess whether the two are associated requires the use of model simulations of the likely climatic effects of the changing atmospheric composition, and the comparison of the results with observations. Three important detection experiments will be discussed here. Wigley and Barnett (1990) used an energy balance climate model (incorporating upwelling and diffusion within the oceans to account for their radiative damping effect). The model was forced from 1765 to 1990 using only the changes in greenhouse gas concentrations, and the response could be varied by changing the value of the climate sensitivity † . In this way, climatic feedbacks, not explicitly modelled, can be incorporated into the model.

The model results were qualitatively consistent with the observations on the century time scale. On shorter time scales, the model failed to reproduce the inter-decadal variability of the instrumental record.

Indeed, this caveat has often been used as an argument against the greenhouse hypothesis altogether. However, Wigley and Barnett point out that such variability represents the background noise against which the greenhouse signal has to be detected. Significantly, the observational record seemed to lie at the low climate sensitivity end of the output range of that predicted by GCMs (1.5 to 4.5°C). However, the situation becomes more complex if other forcing mechanisms, in addition to the enhanced greenhouse effect, are invoked. If the net century time scale effect of non-greenhouse factors (e.g. solar variability, volcanism) involved a warming, the climate sensitivity would be less than 1°C. If their combined effect were a cooling, the sensitivity could be larger than 4°C.

One possible explanation for the decadal time scale discrepancies between the model and observed data is that some other forcing mechanism has been operating which has either offset or reinforced the general warming trend at different times. Using another energy balance model, Kelly and Wigley considered solar variability as a possible candidate. The model was run with a series of sensitivities spanning the accepted range of uncertainty in order to identify the best fit between modelled and observed temperature. Two sets of forcing histories (determined by a 1-D radiative-convective model) were considered, one involving only the effect of enhanced greenhouse gas concentrations (the IPCC 1990 forcing record), the other including also the negative radiative effect of aerosol loading and stratospheric ozone depletion.

Table summarises the results of Kelly and Wigley. The model explicitly calculated the best fit CO_2 doubling temperature (climate sensitivity) and the amount of explained variance in the observational record by each forcing history. As well as IPCC 1990 and 1992 forcing histories, different solar variables (sunspot number, length on sunspot cycle, solar diametre and rate of change of solar diametre) were considered, and combined with the greenhouse forcing.

Table. Summary results of the Wigley/Kelly model

Forcing		Doubling Temp.(°C)	Explained variance (%)		
Green-house	Solar		Green-house	Solar	Total
IPCC 90/92	None	1.8/3.8	46.1/49.2		46.1/49.2
IPCC 90/92	Number	1.2/2.9	39.7/46.1	10.7/6.4	50.4/52.5
IPCC 90/92	Length	0.9/1.9	30.4/33.8	22.6/19.8	53.0/53.6
IPCC 90/92	Diametre	1.1/2.6	37.1/43.6	16.4/10.4	53.5/54.0
IPCC 90/92	Gradient	1.8/3.6	46.1/49.0	8.9/8.0	55.0/57.0

Table demonstrates that a considerable difference in the best-fit climate sensitivities exists between the two IPCC forcing histories. If greenhouse forcing alone is considered the CO2doubling temperature is at the low end of the range predicted by GCMs, and is in agreement with Wigley and Barnett. If the negative forcing of aerosol loading and ozone depletion is also included, the climate sensitivity is much greater. This result is intuitively correct, since the radiative effects of aerosols and ozone loss would offset the warming due to increased greenhouse gases.

When solar variability is included into the model, the explained variance of the observational record is greater than for greenhouse forcing alone. This is true for all of the solar variables considered here. The forcing combination that explains the most variance in the observational record (57 per cent) includes the effects of greenhouse gases, aerosols and ozone depletion, and the rate of change of solar diametre. The latter, it seems, is accounting for much of the inter-decadal variability in the instrumental record.

There now seems to exist a plausible explanation for why the warming has not been regular over the past 100 years, but Kelly and Wigley point out a number of caveats. Most importantly, the explained variance varies little over a range of estimates, about the best-fit value, for the CO_2 doubling temperature. Thus, it remains difficult to define a precise value for the climate sensitivity. Ultimately, knowledge of the climate sensitivity provides the key to projecting the future impact of greenhouse gas emissions. Whilst uncertainty remains, forecasting future climates will remain an imprecise science.

In addition, other possible mechanisms of climate forcing have not been considered. These could include the effects of volcanism, ocean circulation or even the inherent random variability of the climate system. However, as other potential mechanisms are included into the model, the uncertainty in the climate sensitivity rises. If sufficient data existed to define the natural level of climate variability prior to human interference, this estimate could be used to attach error margins to the analysis of the greenhouse signal without explicit consideration of the source of the background noise. Unfortunately, such data do not exit since the period of the instrumental record contains this greenhouse signal.

For the first time, a GCM was able to replicate in broad terms the slow rise in global temperature since the middle of the last century. If greenhouse gases alone were influencing climate, one would expect global temperatures to have risen by some 0.6 to 1.3°C over the last 100 years. By taking into account the anthropogenic sulphate aerosols, the Hadley Centre model simulated a rise in temperature close to the observed 0.5°C.

What makes the experiment so interesting was that the GCM contained simulations of the atmosphere, oceans, ice and vegetation, and can therefore be considered to be a much better representation of the climate system than earlier GCMs, containing only the atmospheric component. Such a model significantly increased the confidence in scientists' assertion that current global warming is due to an anthropogenically enhanced greenhouse effect, albeit muted by the radiative effects of atmospheric aerosol loading. Indeed, it was in response to the Hadley Centre model, and similar ones to it more recently, that the Intergovernmental Panel on Climate Change indicated that the balance of [modelling] evidence suggests a discernible human influence on the global climate does exist. Nevertheless, in view of the reservations highlighted by Kelly and Wigley concerning the uncertainty of the climate sensitivity, it remains arguable whether the cause (enhanced greenhouse forcing) and effect (global warming) have been linked unequivocally.

❐

2

Limits of Global Warming and Environmental Management

Although steps have been taken between Rio 1992 and New York 1997 in the direction of the globalization of environmental protection, the world seems to be further away from sustainability today than it was then. With environmental globalization and the consensual concept of sustainable development, the perception of an ecological crisis, or at least of ecological limits to development, appears to have vanished: acting for environmental protection increasingly tends to be seen as a technical problem, the task of increasing efficiency and of better using resources such as science, technology, information, capital, and institutions. The cause of environmental problems is no longer perceived as linked to industrial development and ever-increasing material accumulation, but it has become the very existence of human beings. Environmental problems are understood as unavoidable, as side effects of human activities, and efforts are then directed at solving these problems.

Indeed, global environmental management and sustainable development can be seen as "problem solving" concepts, in Robert Cox's terminology, as they only represent a strategy to allow the pursuit of present lifestyles and standards.

Following Strange's call for a critical International Political Economy and the need to address the question of "who gets what, how and why," the analysis of the evolution of the system of global environmental management has revealed that it tends to strengthen the mechanisms of exclusion and inequality. Global environmental management and sustainable development tend to be uneven concepts, as they do not aim at promoting the correction of global disparities. They attempt to offer a universal framework in which the global society is the unit of analysis and a large share of the blame for environmental degradation rests on the Third World. Instead of stressing affluence, over- production and over-consumption in advanced countries as the main causes of environmental degradation, it tends to suggest that problems arise from poverty. Environmental degradation is transformed from a problem of affluence into a problem of poverty. The responsibility is shifted from major polluters and industrialized countries' abuses to all inhabitants of the planet. For Chatterjee and Finger (1994), the only different element in this approach is that development is now looked at from a global perspective, making the development discourse universal.

And the New York Summit represented a step further in that direction, asserting the desirability of the globalization process and underlining its beneficial aspects. It also consolidated the role of business and industry as privileged partners of the United Nations, establishing permanent contact and consultation on environmental issues. The regulatory situation relating to TNCs and business in general has worsened greatly in the past five years.

Already in 1992, the US government successfully pressured for downsizing the UN Centre on Transnational Corporations (UNCTC), which had been set up to monitor the social, economic and environmental impacts of corporate investment in developing countries. Today, the UN is considering cutting a sub-group of the UN Human Rights Commission which addresses the impacts of corporations on a broad spectrum on rights issues. The main international initiatives and institution for establishing guidelines for the behaviour of corporations, which could together lay down a code of obligations and rights of TNCs and states, have disappeared.

In their place has come a strong and growing opposite trend to reduce and remove regulations that governments have over corporations, to grant them increased rights and powers, and to reduce the authority of states to control their behaviour and operations.

The Uruguay Round for example has already granted far higher standards of intellectual property rights protection to corporations, thus facilitating further their global monopolization of technology and ability to make profit through higher prices. There are also strong pressures from Northern governments at the World Trade Organization to grant foreign companies the right of entry, establishment and national treatment in all WTO member states.

In addition, the partnership between the UN and global corporations seems to have been further strengthened in the past two years. In part due to the difficult financial and political situation in which the UN finds itself as a result of the US government's refusal to pay the US $1.6 billion it owes, the UN is now openly seeking political and economic support from corporations. At the last Davos Economic Forum in Switzerland, UN Secretary General Kofi Annan called for a human face to the global market and challenged business leaders to adhere to universal values defined by the UN and contribute to global environmental protection, indicating a broader trend of growing UN collaboration with transnational corporations. Recently, the United Nations Development Programme (UNDP) has solicited funds from global corporations with poor records on human rights, labour and the environment, such as Dow Chemical, Citibank and Rio Tinto Plc, in exchange of special UNDP sanctioned logos for use by corporate sponsors. Called the "Global Sustainable Development Facility (GSDF)," the plan calls for corporate sponsors to funnel donations to a separate entity which they will manage. In the words of the UNDP, the GSDF "brings together leading global corporations and the UNDP, to jointly define and implement a new facility to eradicate poverty, create sustainable economic growth and allow the private sector to prosper through the inclusion of two billion new people in the global market economy."

According to the internal memo, sponsors will benefit from the advice and support of UNDP through a special relationship, allowing corporations unprecedented access to UNDP's network of offices, high level governmental contacts and the knock-on effects of its reputation. The plan, revealed through a leaked internal UNDP memo, has been heavily criticized by observers and NGOs, who warn that the interests of global corporations are often at odds with the basic economic and social needs of the world's poor and the values of human rights and environmental protection the UN is meant to protect.

According to Ward Morehouse, President of the US Council on International and Public Affairs, the UN should be monitoring the human rights and environmental impacts of corporations in developing and industrialized nations, not granting special favours... Increasing collaboration will lead to a reluctance to criticize corporations which are central players in the human rights, environmental and developmental dramas unfolding every day across the globe.

To conclude, the UN and international organizations in general seem to be moving towards the adoption of a market-oriented global model of environmental governance, which sees economic globalization as a positive and integrative process. The key actors of economic globalization, transnational corporations, are taking the leading role and consolidating their influence on the system of international environmental governance. With the adoption of this project of global environmental management, one particular understanding of the world, the one promoted by business and large corporations in Western affluent societies, becomes hegemonic and appears to be universal.

Environmental concerns have been incorporated as a mere dimension of the "globalization project," understood in McMichael's definition as "an emerging vision of the world and its resources as a globally organized and managed free trade/free enterprise economy pursued by a largely unaccountable political and economic elite". This project advocates a universalized model of production, of consumption, and thus of dealing with problems of environmental protection resulting from these activities.

By assuming its universality, it tends to marginalize other knowledge and other solutions to problems of environmental protection. Interestingly enough, it was unanimously recognized that the most positive result and follow-up of UNCED was without doubt realized at the micro-level, within the Local Agenda 21 framework. In an effort to implement Agenda 21 locally, social groups have worked together with local authorities to make sustainable development a reality at the local level, often on a truly participatory basis and reflecting grassroots concern and involvement.

From a critical point of view, global management's failure, exemplified at the New York Summit, was not entirely unexpected. Indeed, the global management approach inspired by business perspectives and propagated by the international development establishment tends to strengthen the globalization process at work today, failing to counter its effects in terms of social exclusion and environmental destruction. It tends to weaken social protection and environmental protection in the name of economic efficiency.

It stands at odds with the commitment to social change and to equity that lies at the root of a critical, political economy view of global environmental politics as inserted within the dynamics of economic accumulation and social structures. The question of the ownership of natural resources, for example, is not addressed. However, environmental problems in the South are often linked to problems of resource ownership and equity.

Sustainable Development as defined in Rio and reasserted in New York has been practically translated into technocratic responses to what are in reality political problems.

The purpose of this article was to analyze and evaluate the process of "mainstreaming" of environmental concerns. Recalling the radical and transformative origins of the ecological project, it has provided evidence that environmental concerns have been remodeled by the joint action of technocratic environmentalists, the international UN-related development establishment and business and industry sectors.

Examining the results of international cooperation, the article has questioned the nature of the sustainable development consensus, a consensus deeply marked by the growing access and influence of global corporations on UN activities. Today, market-oriented perspectives to environmental problems seem to be prevailing over more transformative views, especially at the international level, within the framework of international organizations and institutional agreements.

The article has suggested that the mainstream approach to sustainable development tends to reduce ecology to a set of managerial practices aiming at resource efficiency and risk management. In doing so, it tends to address a civilizational impasse as a mere technical problem. The mainstream approach proposes that environmentalists should operate using the language and the worldview of Western economics in approaching ecological concerns.

Instead of designing cultural and political limits to development, the project of "global environmental management" tends to become part of a technocratic effort to sustain industrial development in the age of economic globalization. Environmental protection, together with democracy, human rights and free market economics, becomes a universal consensus, a universal consensus which, as Baudrillard remarks, arouses suspicion, since it is about values that have become devalued, values becoming emptied at the very moment of their hegemony. Environmental concerns become just another element in a process leading to global uniformity, a uniformity of cultures, lifestyles, mentalities, but also of relationships with nature.

This market-oriented agenda may provide a starting point for dealing with global environmental problems.

The documents, which emerged from international environmental negotiations from Rio to New York, replete with inconsistencies, represent a complex mix of disagreements, hopes and compromises. One may concede that conventions and obligations reflect the need for government negotiators to find the minimum agreeable grounds to initiate a large open-ended process on major environmental issues.

In this sense, they only produced a general framework for negotiations, steps on the way to building international regimes. They do not form a series of real commitments representing an effective consensus on how to deal with global environmental issues.

Yet considering the amount of time, energy and resources invested in this process of international environmental regime building, one might have hoped for more concrete, positive results. The failure of the present framework to effectively promote sustainability, which became evident in the 1997 New York summit, is recognized even by one of the major promoters of this path, Maurice Strong, ex-secretary general of

UNCED, today President of the Earth Council. For Strong, unfortunately, the economic, social and demographical forces that lead to unsustainable development still prevail. Strong sees the lack of political will from governments as the main cause of this failure.

The present framework appears to contain many contradictions that limit the ability of the international and national communities to solve satisfactorily environmental problems.

Adopting an international political economy perspective, this article has argued that the main problem that international efforts to protect the environment have to address is the issue of the impact of economic globalization.

Economic globalization has in a sense helped to create conditions for the development of policy mechanisms and institutions that will universalize and promote the concept of "sustainable development." Global change is exerting a structuring influence on the redefinition of environmental politics.

However, the kind of sustainable development being promoted seems to represent more the consolidation of a global project of "environmental management" than a real shift away from destructive practices. Globalization is consolidating a market-friendly view of sustainable development, a view that gives priority to the sustainability of "global growth" and to the correction of environmental damage. This tends to be carried out at the expense of the competing alternatives and participative view of sustainable development as stressing not only development but also social equity and decentralized participation.

The "globalization project" has shaped and redefined both the content of environmentalism and environmental policies and structured the international political economy in a way that makes sustainability more difficult to achieve.

GLOBAL WARMING SCENARIOS

Although the science remains uncertain, 1990 marked the year when many governments accepted the impending reality of global climate change. Some governments began to announce targets for reductions in carbon dioxide in order to prompt the international negotiation process. Some governments remain unconvinced that the greenhouse effect is real. Others accept the science, but not the need to act dramatically. This volume is concerned not with the scientific debate as such, but with the basis for climate change policy in OECD countries. Crucial to the whole process of policy development is an assessment of the impacts of climate change.

These impacts will appear first as physical changes in the environment, but will eventually translate into social and economic changes as well. Impacts are important, essentially because they represent the potential benefits that might be realised from taking political action against the climate change problem.

If impacts are projected to be small, or if it is likely that the forecasted impacts could be easily managed, then the benefits of measures to contain global warming will be small, and the balance of costs and benefits will militate against major policy action. On the other hand, if the impacts are projected to be large, or if there is a real risk of some of them taking on "catastrophic" proportions, then early and firm action would be prudent. Put another way, the cost-benefit balance, however imperfectly understood, is fundamental to the policy process. Because the costs of environmental policy are typically easier to quantify than the benefits, policy discussions often begin with an exogenously-determined "target", and seek to minimise the costs of achieving that target. This is more cost-effectiveness analysis than it is cost-benefit analysis. It has the advantage of being conceptually simpler, but it runs the risk that the initial target will be inappropriately set. If this target is too stringent (or not stringent enough), there is a likelihood that both economic and environmental resources will be "wasted".

Given the long time horizons involved in the climate change issue, and given the world-wide economic and environmental implications of that issue, the potential for such wastage is very high indeed. It is a basic premise of this book that economic resources should be allocated to resolving the climate change problem (i.e. costs should be incurred) only to the extent that the benefits (i.e. reduced damages potential) warrant these costs. In order to define this "tradeoff" point, an effort has to be made to measure the potential impacts of global warming.

The Intergovernmental Panel on Climate Change (IPCC) was set up by the United Nations Environment Programme (UNEP) and the World Meteorological Organisation (WMO) in 1988 to investigate global warming. It established three working groups to help it fulfil its mandate. Working Group 1 (Chaired by the United Kingdom) has looked at the scientific evidence in support of global warming.

Working Group 2 (chaired by the Soviet Union) has investigated potential impacts. Working Group 3 (chaired by the United States) has evaluated various policy response options.

Working Group 1 has forecast that global warming will indeed occur, and has assessed four possible scenarios:

Scenario A: "Business as usual" i.e. a continuation of current trends in greenhouse gas emissions, which will produce a rate of warming in terms of global mean surface temperatures of 0.2°C-0.5°C per decade in the 21st century, with a "best guess" of 0.3°C. This would produce a 1 degree rise in temperature by 2025 and 3 degrees by 2100, compared to 1990 levels.

Scenario B: Deforestation is halted, natural gas is increasingly substituted for coal, energy conservation occurs, but temperatures still rise by 2°C by 2100 compared to today, i.e. a rise of 0.2°C per decade.

Scenario C: A greater switch to renewable energy sources in the second half of the 21st century holds the temperature rise to a little above 0.1°C per decade.

Scenario D: Assumes that the switch to renewables occurs in the first half of the 21st century, which stabilises gas concentrations in the atmosphere.

The temperatures in question are global mean temperatures. For policy purposes, these estimates are not very useful. What matters more is the regional variation which will exist around about the mean. For example, in the "business as usual" scenario, temperatures are expected to be higher than the mean in Southern Europe and central North America, with reduced summer rainfall and soil moisture.

By 2030, winter temperatures could rise in these regions by 2°C (perhaps more in Central North America), and summer temperatures by 2-3°C. In the Sahel, warming will increase by 1-3°C. South East Asia might experience temperature rises of 1-2°C, and Australia some 1-2°C in summer and 2°C in winter.

Sea levels rise in the "business as usual" scenario by some 6 cms per decade due to thermal expansion of the oceans, and the melting of some land ice, producing a rise of some 20 cms by 2030, and 65 cms by 2100.

Regional Shares of Emissions

The importance of the USA and USSR/Eastern Europe are clearly seen, as is the role of European Community countries (which make up the bulk of the "rest of OECD"). The rapid growth of the developing countries' shares is also seen - note the very rapid rise in the share of Centrally Planned Asia (China mainly), and other developing countries. Any agreement on emission controls must, therefore, secure the co-operation of both developed and developing countries, if it is to be successful.

Impacts

The Working Group 2 report emphasises the fact that uncertainty is a major problem to be overcome in assessing the socio-economic impacts of climate change. Given the large uncertainties that are involved, it is understandable that most of the impacts described in their report are outlined in qualitative terms only.

Nevertheless, Working Group 2 foresees significant impacts in virtually all economic sectors, assuming that temperatures and sea levels rise as predicted. More specifically, Working Group 2 recommends that efforts should be intensified to better understand the linkages between physical and socio- economic impacts, and that better methodologies for quantifying these impacts need to be developed. To some extent, this conclusion is based on the three methodological studies contained in this book. Working Group 2 recognised early in its mandate that hard data on the impacts of climate change simply does not exist.

Some studies had been carried out on potential physical impacts, but virtually no guidance existed on the possible "downstream" social, and economic response functions. For that reason, Working Group 2 asked OECD to work on developing "more robust" methodologies for assessing these types of impact. The papers presented in this book are a direct response to this request.

Table: *Regional contributions to Greenhouse Gas Emissions*

CO_2 Anthropogenic Emissions Shares (%)				
Source	**1985**	**2000**	**2015**	**2050**
United States	21	19	16	12
Rest of OECD	22	19	16	12
USSR and Eastern Europe	22	22	19	18
Centrally Planned Asia	10	13	16	21
Other Developing Countries	25	28	32	37
Commercial Energy	86	87	89	92
Tropical Deforestation	12	11	9	6
Other	2	2	2	2
Total (109 tonnes Carbon)	5.99	8.05	10.27	16.95
Average Annual Growth Rate	1.6%			
Methane Anthropogenic Emissions Shares (%)				
Source	1985	2000	2015	2050
United States	12	11	9	8
Rest of OECD	13	12	12	10
USSR and Eastern Europe	13	14	14	15
Other Developing Countries	17	16	17	19
Other Developing Countries	46	47	49	48
Fuel Production	18	22	26	32
Enteric Fermentation	23	24	23	22
Rice Cultivation	34	31	29	24
Landfills	9	10	10	14

Tropical Deforestation	6	6	5	4
Other	9	7	7	
Total (101 tonnes CH_4)	320.1	399.5	476.8	710.5
Average Annual Growth Rate	1.2%			

Policy Options

How should the international community respond to the issue of global warming? Working Group 3 points out that there are basically three options:

Do Nothing

The arguments for doing nothing are:

— That global warming is all very uncertain anyway;

— That doing something may be expensive and that costs borne today will benefit generations yet to come (an inequitable situation);

— That delaying present action will improve the level of scientific information about global warming; and

— That the world will adapt "automatically" through migration, changing agricultural production, switching expenditures to sea level defences, and so on.

The arguments against doing nothing are:

— That global warming is irreversible. The longer the delay, the more warming the world is committed to. If the worst fears are realised, it will be too late to correct it;

— That scientific information can be improved even while further action is being considered;

— That delay is "intergenerationally unfair"—current generations have an obligation not to impose heavy costs on future generations; and

— That action now could be relatively cheap (e.g. energy conservation), and will likely yield other environmental benefits (e.g. reduced acid rain; reduced congestion, if the contribution of road travel to CO_2 is properly tackled, etc.).

Reduce Emissions and Invest in Carbon "Sinks"

The arguments for reducing emissions are very much the same as the arguments against doing nothing. The additional advantage of acting to reduce the problem at source is that it gives greater certainty compared to adaptation policies. Most importantly, global warming may have a large potential for "surprise" events, e.g. major climatic incidents, simply because we know so little about how climatic systems function. These surprises may be very damaging. By reducing emissions, we lower the potential for unpleasant surprises.

The obvious argument against reducing greenhouse gas (GHG) emissions is that, while it may be fairly cheap in the short run, further reductions in emissions will get more and more expensive. A number of studies have estimated costs of adjustment, and some have suggested that major reductions in GHGs could be very costly in economic terms.

Afforestation offers the potential for "fixing" CO2, although due allowance has to be made for carbon release once the forest rotation has ended and the wood decays. Much depends on how the wood is used. For example, wood for paper pulp would release carbon fairly quickly, or would contribute to methane releases in landfill sites. Wood used for furniture and building could store carbon for much longer. Afforestation in the tropics might fix carbon on an even bigger scale.

Afforestation also introduces the idea of "carbon offsets". An electric power utility might be allowed to build a new power station, provided it "offset" the projected emissions with carbon-fixing trees. Since it does not matter where the emissions and fixing take place (in terms of global warming), the prospect is opened up for tropical afforestation as an offset to increased carbon emissions in the rich world. This is already happening. One US utility has agreed to conserve tropical forest in Guatemala. Dutch utilities may well plant new forests in Colombia, Ecuador and Peru. The arguments against afforestation as a "cure" are:

- It could be very expensive;
- Much modern afforestation is regarded as being environmentally unattractive anyway (e.g. conifer plantations); and
- In order to overcome the problem of merely postponing carbon releases (because of wood decay), afforestation would have to proceed at a faster and faster pace over time.

Invest in Adaptation

Under the "do nothing" option, adaptation is "natural": it simply comes about because people respond to climatic change. But governments could also invest in adaptation by e.g. raising existing sea defences and building new ones; investing in new research and development (e.g. more climate resistant crops); protecting groundwater systems from satinisation; relocating communities; etc. The argument for adaptation is that it may be a lot cheaper than emission reduction, especially if costs could be spread out over very long periods of time.

The argument against adaptation is that it does nothing to reduce further warming, and simply imposes bigger and bigger cost burdens on future generations. Although it is illustrative to fix our attention on certain threshold points in time (e.g. the year in which CO2 emission levels are expected to reach twice their current magnitude), it should not be forgotten that warming will continue long beyond this point, if no action is taken.

Choosing a Policy Mix

In the end, international global warming policy will probably consist of some measure of all three of these options. In particular:

— Some warming will be allowed to take place, simply because the economic dislocation caused by aiming for stabilisation of temperature is likely to be too great;

— Because some warming will take place, both "natural" and "engineered" adaptations will occur;

— The international community is already focusing on emission reduction (coupled with afforestation) as its primary policy response. From the discussion above, it is clear that, so long as the welfare of future generations matters, this is a sensible approach.

ENVIRONMENTAL MANAGEMENT

Environmental management is not, as the phrase could suggest, the management of the environment as such, but rather the management of interaction by the modern human societies with, and impact upon the environment. The three main issues that affect managers are those involving politics (networking), programmes (projects), and resources (money, facilities, etc.). The need for environmental management can be viewed from a variety of perspectives. A more common philosophy and impetus behind environmental management is the concept of carrying capacity. Simply put, carrying capacity refers to the maximum number of organisms a particular resource can sustain. The concept of carrying capacity, whilst understood by many cultures over history, has its roots in Malthusian theory. Environmental management is therefore not the conservation of the environment solely for the environment's sake, but rather the conservation of the environment for humankind's sake. This element of sustainable exploitation, getting the most out of natural assets, is visible in the EU Water Framework Directive.

Environmental management involves the management of all components of the bio-physical environment, both living (biotic) and non-living (abiotic). This is due to the interconnected and network of relationships amongst all living species and their habitats. The environment also involves the relationships of the human environment, such as the social, cultural and economic environment with the bio-physical environment. As with all management functions, effective management tools, standards and systems are required. An 'environmental management standard or system or protocol attempts to reduce environmental impact as measured by some objective criteria. The ISO 14001 standard is the most widely used standard for environmental risk management and is closely aligned to the European Eco-Management and Audit Scheme (EMAS). As a common auditing standard, the ISO 19011 standard explains how to combine this with quality management. Other environmental management systems (EMS) tend to be based on the ISO 14001 standard and many extend it in various ways:

- The Green Dragon Environmental Management Standard is a five level EMS designed for smaller organisations for whom ISO 14001 may be too onerous and for larger organisations who wish to implement ISO 14001 in a more manageable step-by-step approach
- BS 8555 is a phased standard that can help smaller companies move to ISO 14001 in six manageable steps
- The Natural Step focuses on basic sustainability criteria and helps focus engineering on reducing use of materials or energy use that is unsustainable in the long term
- Natural Capitalism advises using accounting reform and a general biomimicry and industrial ecology approach to do the same thing
- US Environmental Protection Agency has many further terms and standards that it defines as appropriate to large-scale EMS.
- The UN and World Bank has encouraged adopting a "natural capital" measurement and management framework.
- The European Union Eco-Management and Audit Scheme (EMAS). Other strategies exist that rely on making simple distinctions rather than building top-down management "systems" using performance audits and full cost accounting.

For instance, Ecological Intelligent Design divides products into consumables, service products or durables and unsaleables-toxic products that no one should buy, or in many cases, do not realize they are buying. By eliminating the unsaleables from the comprehensive outcome of any purchase, better environmental management is achieved without "systems". "Today's businesses must comply with many Federal, State and local environmental laws, rules, and regulations. It's vital to safeguard your company against compliance shortcuts. This approach leaves you vulnerable to violations of the law, in addition to missing important environmental liabilities. "Who Needs Environmental Management?

All businesses need a sound environmental policy, and effective implementation. Certain business sectors and activities have been mandated to have effective environmental management practices, limits on emissions, prevention, and emergency response systems. Worldwide as of April 2005, 88,800 organizations have achieved compliance and registration to the ISO 14001 environmental management standard. Japan has the most registrations, followed by China. USA is ranked 6th, and Canada is 12th. The number of ISO 14001 registrations per $ GDP or population are interesting statistics. In 2001, the USA had one registration per 200,000 people or one per $6000 million of gross domestic product (GDP). This compares to one registration per 23,000 people in Japan and one registration per $900 million of GDP in the United Kingdom. For Canada as of August 1999, it is 34th in the world in terms of ISO 14001 certifications per unit of GDP, behind all major competitors, and behind such countries as Korea, Thailand, Brazil and Mexico. Clearly, North America is lagging behind the rest of the world in adopting ISO 14001. EMAS, the European environmental standard, has achieved similar recognition and adoption. The rate of growth in ISO 14001 registrations has been exponential since its start in Dec. 1995. Individuals also need awareness and limits on their polluting behaviours that affect others.

Dimensions of Environment Management

A fruitful approach to studying environmental management is the analysis of industry-government relationships. Policy is defined and translated into action in this context. We find that industrial managers legitimize government regulation of industrial water use.

However, the pattern of responses suggests that some of the support may be based on economic self-interest and industry's ability to control agency action. Managers legitimize a policy-setting role for federal agencies and an enforcement role for local and state government. Across the Asian region, natural resource exploitation is accelerating dramatically as countries, cities and small communities are ever more incorporated into the global economy.

Economic reform programmes that favour domestic and global market expansion rather than a social welfare agenda, policy responses to climate change, pressures associated with population growth and intensified geographical mobility, and urbanization and commoditization, are reconfiguring patterns of natural resource use and governance at both a national and local level and are having complex effects on peoples' lives.

These processes are themselves not innocent of gendered power relations: they are inflected with gender discourses that set in motion differentiated and unjust life opportunities and exclusions.

At the same time, sustainable development policy initiatives that seek to ameliorate environmental degradation and its negative livelihood effects not only bring gendered impacts and responses, they also work through and produce particular framings of gender and gendered power relations. The impact of this is apparent in the unintended consequences associated with sustainable development initiatives that target women as a homogeneous and undifferentiated social category, at times exacerbating social and gender injustices.

Related to this process of globalized marketization, natural resource management is embedded in the increasing diversification of livelihoods: a process buoyed by policies that support entrepreneurialism and appear to be producing individualized portfolio livelihood strategies across Asia.

The Challenge Of Environmental Management

Before moving on to introduce the elements of management practice and strategy which will facilitate improved environmental performance in industry, it is important to relay some of the factors which together serve to encourage industry to respond to the environmental challenge.

Environmental Efficiency

Companies often strive to minimize the costs of their operations. This is especially relevant in relation to the efficiency with which they use their material inputs. As the ability of the environment to supply raw materials and accept waste is diminished, the costs of these services to industry will increase. As a result, more efficient raw material utilization and a decrease in the amount of waste generated are key factors which will encourage industry to minimize its environmental impact. Particularly in relation to waste products, companies are experiencing increasingly stringent legislation which increases the costs of waste management. Waste should be viewed both in terms of physical waste generated and the less tangible losses experienced through an inefficient use of resources. Avoiding these losses improves both the business and environmental performance of a company. As a result, many companies have pursued a strategy of waste minimization for a number of years and have experienced short payback periods on investment in waste management. In efforts to increase the efficiency of their operations, many companies have developed integrated management systems to reduce inefficiencies and the likelihood of errors. Most commonly to date, these have centered around the promotion of quality.

The Influence of Government

The main impact of government on the environmental performance of industry has been through the development of environmental legislation. Environmental considerations have been built into the legislative framework for many years. Initially, establishing rights of ownership over natural resources led to the development of a legal system to protect those rights. Subsequently, the impact of industrial activity on the health of employees and the surrounding community kd to the creation of public health and safety legislation. Measures have also been introduced to control the use of products, processes and wastes which may harm the environment. The impact of environmental legislation on the operation of industry has been profound and is set to become ever more tough. As the strain placed upon the environment mounts and knowledge of the causes and effects of environmental degradation becomes more complete, the extent and impact of environmental legislation will continue to develop. Thus, industry must satisfy an increasing number of legal obligations in relation to the effect that its activities have upon the environment. As a result, in all of its operations, industry must plan ahead to meet the demands of current and forthcoming environmental legislation. By developing proactive responses to legislative pressure, industry will reduce its costs and exposure to risk.

While in the short term, legal obligations undoubtedly increase the costs of production that fall upon the firm, it is up to each firm to comply with legislation in the most cost-effective way. The development of proactive strategic responses to the demands of legislation will reduce these costs. In parallel with the development of environmental legislation, governments are increasingly applying market instruments to achieve environmental objectives. Actions of this nature may include the imposition of taxes on environmentally damaging goods, subsidies on environmentally friendly goods or the provision of information relating to the environmental performance of companies or products. Market instruments are intended to channel the choice of consumers or other stakeholders towards the better environmental option. Thus through a combination of legislative and market instruments, by encouraging certain activities and discouraging others, governments seek to accelerate the structural change which encourages improved environmental efficiency in the economy as a whole.

The Development of Stakeholder Influence

Individual businesses interact with a number of stakeholders, all of whom have an interest in the performance of that company. Traditionally the main focus of stakeholder interest has been upon the financial performance of the company. Increasingly, however, stakeholder pressure is concentrating on the environmental performance of the company. The range of stakeholders which demand high environmental standards.

Customers

The relationship between a company and its customers is obviously of para-mount importance. In relation to environmental considerations, the potential importance of green consumerism cannot be overstated. The range of characteristics that underlay the purchasing decision are a fundamental consideration for all businesses. Increasingly, the environment is being accepted as one such characteristic by consumers. At present, however, the influence of green consumerism on most businesses is marginal. Of the myriad of products that each consumer buys, very few are chosen on the basis of their environmental credentials alone. Nevertheless, it is certain that credible claims relating to environmental performance constitute one positive element among the many characteristics upon which consumers base their purchasing decision. Companies which can validate and communicate the environmental performance of their products will enhance their competitive position. Governments are also seeking to increase the potency of green consumerism by providing the consumer with the information necessary to make an informed choice in relation to the environmental performance of each product within the product range. For this reason for example, we saw the introduction of the EC's eco-labelling scheme in 1993.

Trading Partners

Many businesses do not sell into 'end-consumer' markets and may therefore perceive themselves to be remote from any consumer pressures to improve their environmental performance. Increasingly, the pressure to improve environmental performance is emanating from trading partners rather than the ultimate consumer. In efforts to improve overall environmental performance, many companies are exercising their own rights both as purchasers and as vendors and are demanding that all of the companies within their supply chain seek to minimize their own environmental impacts. Hence, demands to improve environmental performance at all stages in the supply chain are being diffused beyond those companies that are directly exposed to the pressures of green consumerism. An increasing number of companies are preferring to buy their resources from or sell their products to companies which meet certain standards of environmental performance.

The provision of information on company environmental performance through standards such as BS7750, the British Standard on Environmental Management Systems, and the EC's eco-management and audit scheme will increasingly be written into contracts in the future. Increasing environmental concern and the improved provision of information relating to environmental performance will reward those companies which achieve and communicate high environmental standards with a competitive advantage.

The Community

Industry shares its surrounding environment with the local population. Increasingly this population is demanding a high level of environmental performance from its industrial neighbours, and seeks some degree of reassurance that they are not exposed to significant environmental risk due to a company's operations. This concern has been recognized for many years and was initially recognized in public health legislation. Trends towards freedom of access to environmental information will give greater power to local communities when they question the activities of local industrial co- habitants. In order to foster a positive working relationship, companies must improve their environmental performance and communicate their efforts to the surrounding communities. This is true both for future developments and existing operations.

Employees

The population in the community surrounding a company also includes the workforce of that company. The pressure to provide a healthy living environment is magnified within the workplace. Employees seek healthy and secure working conditions, and can draw on an established framework of health and safety legislation in this respect. However, employees' concerns relating to the environmental performance of their employers goes beyond the impact of operations on the working and living environment. Increasingly people wish to work for ethical and responsible companies. Companies that reflect the environmental concerns of the public will find it easier to attract, retain and motivate a quality workforce.

Investors and Insurers

The pressures to improve environmental performance also emanate from the investors and shareholders of a company. The rapid growth of ethical to lend their financial support to companies which behave in a responsible manner. There are also a number of very good business reasons why investors prefer to work with companies that have a proven track record of environmental integrity. The structure of legal liability for environmental damage dictates that any party that causes environmental damage may be fined and required to bear the costs of remediating that damage and to compensate the affected parties for any associated losses. It is increasingly difficult and expensive to obtain insurance to cover such issues. Consequently, companies associated with a significant environmental incident may suffer significant financial losses. These losses are then translated into reductions in the share price and the associated dividends.

Banks that lend to companies secure the loans on the basis of the physical assets of the company and often on the land upon which any investment takes place. Should the company cease to be viable, the bank assumes ownership of those assets which are then sold to cover any outstanding debts. However, should the physical assets of the company be contaminated, then the value of the assets is significantly reduced. Indeed, the banks may inherit any environmental liabilities that the liquidated company generated. Commercial lenders are therefore reluctant to lend money to any company which may develop any environmental liabilities or to secure loans on the value of an asset which may be eroded through contamination. As a result, companies which cannot demonstrate a high level of environmental performance associated with low environmental risks will find it increasingly difficult and expensive to attract and retain investment and insurance for their operations.

Media and Pressure Groups

A combination of increased public awareness of environmental issues and freedom of access to information on the environmental performance of companies will serve to magnify media and pressure group interest in the environmental performance of industry. In order to manage media and pressure group attention, companies must be able to state that they have made efforts to reduce their environmental impact. However, while it may be tempting to allow the PR or marketing departments to lead the way in convincing all stake-holders of this commitment, any shallow or spurious claims will soon be uncovered. Claims which cannot be substantiated are likely to be seized upon and will be very detrimental to a company's public image. Companies which seek to communicate responsible environmental performance must base any claims that they make to this effect on hard facts which they are willing to communicate.

BROADENING ENVIRONMENTAL MANAGEMENT

Environmental management should be considered as a means for attaining sustainable development. However, the direction of this development should not be solely determined by a purely ecological environment, but must include the human aspect, as expressed by cultural diversity. Although present-day indigenous populations may still display original characteristics that illustrate their unique ways of adapting to the natural as well as to the human environment, often the most authentic expressions of knowledge and know-how are hardly remembered and lie hidden in the archaeological heritage. These remains of the past may hold a huge potential for understanding how we became what we are and what lessons past successes and failures may teach us, but they are, at the same time, prone to destruction for multiple reasons.

Current protection measures don't work well, partly because of a lack of efficient enforcement, and partly because of the great complexity of the problem. All development projects in archaeologically interesting areas should therefore include an archaeological impact assessment (AIA), and foresee appropriate measures for protecting ancient remains. This means that political will, at all levels, must exist to include AIA in the project cycle and to execute it properly, after which monitoring of the archaeological heritage must be organised. Hence the need for awareness-raising and capacity-building at all levels. A possible way of reaching this objective could be to include the cultural dimension in existing legally binding environmental agreements on the international level, or to negotiate a convention sui generis that combines both aspects. If sustainable development means that we can continue to use all kinds of resources for our own development, while guaranteeing that future generations can also establish and enjoy a good quality of life, and if, furthermore, we accept that sustainability comprises aspects of natural environment, social welfare and economic well-being, then the very concept of sustainable development must be broad enough to encompass the multiple facets of human society.

The omnipresent, though mostly unconsidered, cultural characteristics of human society in general make up a very important aspect of the driving forces that ultimately determine the direction of development. However, human society varies widely according to historical background and geographical setting. Hence the need to take into account the cultural specifics of countries, or regions, when designing development plans.

Regardless of their diversity, cultures share general components (social, economic, political, and religious) which relate to their earlier stages of development or to foreign inputs, and show the ways in which they managed to interact with, and adapt to, their environment, both natural and cultural. Many clues to this understanding can be found in the cultural heritage in general, but particular aspects may actually lie buried in the archaeological heritage.

Since this book focuses on the practice of evironmental management and the cultural dimension therein is not currently taken into account, we shall divide this into a first part that gives some basic definitions, a second section with more practical points about cultural and archaeological heritage, environment and development, including examples in the form of case studies, and a final part suggesting some ideas for the future. Since the author's experience in this field mainly lies in South America, the topic under discussion is written against this background and with examples from Ecuador. When transposing the same problems to other regions, some degree of adjustment may be necessary.

First of all, we consider environmental management to be a means for sustainable development, not an end in itself. Sustainable development must be the objective for human society as a whole, be it in the developed world or in the developing world. The first needs to become a healthier society, while the latter must catch up in a sound way. As stated in the

Brundtland report: Sustainable development is not a fixed state of harmony, but rather a process of change in which the exploitation of resources, the direction of investments, the orientation of technological development, and institutional change are made consistent with future as well as present needs.

This means that, for the time being, we must limit the use of the environment if we don't want to exhaust our natural resources. Simultaneously, we need to help poor countries develop in a sound way, which, in turn, demands an extra amount of resources. In order to resolve the apparent contradiction, we must combine environmental policy with development policy. In doing so, we should be able to reduce poverty, increase democracy and control consumption and demography.

However, care must be taken not to establish a one-sided, narrow-minded concept of sustainable development, which would then tend to impose itself in a rigid way everywhere. It would almost certainly be a western point of view, based on exclusively western assumptions about development and sustainability. This would only be a repetition of past errors and be doomed for quick failure. On the contrary, a lot of input is needed from the quiet majority, the ones that traditionally have been recipients and have been directed towards targets that were not their own. If we are serious about sustainable development, we need to accept that there is not one but several facets to sustainability, reflecting the many actors who will be participating in the drama of the future.

Culture and Sustainable Development

When speaking about development, culture usually is not an issue. When considering development assistance, cultural aspects are hardly ever included in plans, programmes or projects. A possible explanation is that development assistance has always been regarded as a purely technical matter, while culture has been viewed rather in the opposite way. A more important and, at the same time, more negative distinction between both has to do with feelings of solidarity with broad population levels when talking about development assistance while thinking in terms of distance, abstraction and being somehow elite-oriented when mentioning culture.

Culture can be defined as learned behaviour, transmitted from generation to generation, for coping with both the natural and the human environment. It is therefore based on tradition, but also on innovation, and is characteristic of a geographical region at a certain point in time. Cultures can be very long- lived, evolving slowly in one place, or expanding their influence in an imperialistic way. They can disappear suddenly, due to disaster, or be conquered and replaced by other human groups. Wherever cultures are located and whenever they existed, the natural environment has always been of key importance, by the mere fact that human development has been a process of progressive domination of nature in order to ensure survival. They have been managing their environment in their own right, in the same way as we are trying to do now. In this continuum, our ideas may replace earlier ones or combine past and present knowledge.

Although culture dynamics can ultimately teach us a lot about the success stories or failures of extant or extinct social systems, the careful observation of particular culture traits may give insights into their unique ways of adapting to specific environments. These ways often include ideal solutions to particular problems, examples of adapted technology that are generally more environment-friendly than present-day practices.

As states, indigenous people often have a unique knowledge of their environment and its natural characteristics. In combination with modern scientific research, this knowledge can produce new opportunities for sustainable development. Therefore, indigenous people should be given more rights of self-determination, should administer their own natural resources and participate in the decision-making. Finally, indigenous people should be protected against dominant societies that threaten their natural resources, culture, prosperity and way of life.

First, the many different, specific cultural answers to environmental conditions can form an inventory of possible alternatives for good environmental practices in development policy. Therefore, thought should always be given to local knowledge when planning development actions, seeking direct involvement of the local population. In this regard, recent trends in development assistance and environment, calling for more bottom-up and participatory approaches, must be warmly welcomed.

Second, it will soon become clear that not only cultural specifics, that are technical solutions, may play an important role in finding the right solutions, but the whole of society targeted by the development activities must have an input into the process. Only then can development be socially integrated, guaranteeing not only positive direct results, but also long-term improvements and multiplicator effects.

In the long run, repeating the exercise over and over again will enhance cultural sensitivity on both sides of the development co-operation, making for better results, closer relations and mutual understanding of each other's needs and deeds.

Cultural Heritage

Cultural heritage encompasses all possible expressions of a society. We generally think of material phenomena such as art, technology, architecture, etc., but whatever particular cultural trait we look at, no abstraction can be made of the ideas and beliefs embedded in it. One of the main problems in the present-day cultural reality of certain societies is the degree to which the original traits have been altered by the importing of foreign ('modern and progress-oriented') elements, such as technology, economic systems, models and the like. A good example is South America, where Iberian conquest willingly changed all aspects of pre-Columbian society. By now, 500 years later, a lot of knowledge would have been lost forever, were it not that historical and archaeological research is recuperating it.

The developed countries also have an important share of the responsibility in the cultural realm. Haven't we imposed our culture on many societies in different latitudes, thereby inflicting serious damage or actually destroying unique local cultures? The principle of 'the polluter pays' may work in the material dimension, provided there is still a way to mend the damage and that nature can regenerate, but on the cultural level, being entirely man-made, the destruction may have been terminal. If a local population has been exterminated, who is going to bring it back to life? At most, it can be re-created to serve as a showcase in some ethnographic museum, but the living thing, the dynamic society, has stopped existing. Need it be said that the developed world was a cultural depredator long before turning to destroying nature? If the origins of the latter can be tracked back to the industrial revolution, did not the first start shortly after the Neo-lithic revolution, several millennia earlier?

As rich countries, with the extra resources and knowledge to take care of cultural heritage, it is our duty to be guardians not just of our own cultural heritage and that of the countries we are related to, but also of world heritage, so as to ensure that future generations can know and understand the rich cultural diversity on our planet. In order to do so we must decisively undertake actions to preserve and protect the cultural heritage. This means that we should invest in the safeguarding of culture, for we can only protect and preserve those cultural elements we know. Through anthropological research in the field, a lot can be learned about important aspects such as local political and social organisation. Present-day knowledge and technology can be studied through direct contact with people, but certain traditional features may be harder to detect by simple observation or interview. Some of them may actually be forgotten, buried in the minds and in the soil. Archaeological research may therefore not only recover the origins of the local population, explain the reason for their presence and actual situation, but additionally discover fine examples of ecologically well-adapted technology, dating from centuries ago.

Several other examples could be mentioned, but entering into the details would lead us too far. There are, however, interesting elements on the level of infrastructure, such as condensing water from daily mist and using it for irrigation in the desert west coast of South America, or increasing agricultural areas and preventing soil erosion by a wide array of terracing and channelling techniques, of which many good examples exist all over the world. On a smaller scale, and still awaiting further exploration, ancient technology continually startles researchers with many examples of simple and very logical technical solutions for a wide variety of problems in metallurgy, ceramics, architecture, food storage, etc. They may not serve industrial purposes, but they can still be applicable on the artisan level.

Recuperating Technology

In many regions of South America, a traditional agricultural system known as 'raised fields' has been reported since ancient times. It is composed of raised parcels of land, limited by small, shallow channels, which are interconnected and feed into bigger watercourses, while being interspersed with dammed areas for stocking water reserves. This way a dynamic system of drainage and irrigation is created. When located in tropical lowlands, this is its main function. However, identical agricultural systems also occur in the interandean valleys or high altitude plains of South America.

Here, research has proven that during the dry period in the Andes the water in the channels creates a microclimate by absorbing sun heat during the day that is radiated at night, providing the extra degree or tenths of a degree Celsius that may make the difference between conserving the crop or losing it all by freezing. As these raised fields are located on the valley floor or lower slopes, drainage/ irrigation may also be a function.

Archaeological investigations of the raised field complex to the North of Guayaquil, Ecuador, have dated the earliest phase of this system as far back in time as 2000 BC, increasing its extension and complexity over time. During the last pre-Columbian period, that is AD 500-1500, vast areas of the Guayas plain were covered with raised fields, articulated around a hierarchy of settlements, providing the economic basis for chiefdoms that reigned over the tropical wetlands behind the pacific coastline.

Not only did the raised fields yield different kinds of crops, planted higher or lower on the slope of the field according to the need for more or less water, but the channels too provided an important part of the diet, through the fish and molluscs they contained, and the birds they attracted. Maintenance of the system required periodic cleaning of the channels by removing the deposits and spreading them on the fields, as a natural fertiliser. This way the raised field system became an interestingly diversified source for subsistence and surplus production.

Excavations of raised fields and settlements associated with them have given a clear picture of how adapted technology could sustain a large population and its social development for many centuries, failing only at the moment when the Spanish Conquest disrupted the indigenous society. Experiments with planting crops on selected fields have proven the potential of the system even today—an excellent example of lessons to be learned from the past and possibilities for applying them to the future.

Archaeology

Archaeology is the scientific discipline that studies past cultures, and their development, based on the material remains that these cultures have left behind. It does so through surveying the landscape, registering all kinds of visible remains and by excavating sites where those remains are buried. The remains of the past are not only limited to what we traditionally consider as such, that is objects and buildings, but include all things produced, used or slightly modified by man, comprising a variety of elements—bone, seeds, soils, stone, wood, shell, metal, ceramics, etc.

All these elements, together with the specific context in which they are encountered, and its structure, allow the archaeologist to build a sequence of cultural events, reconstruct the way of life at different points in time and infer the broad historical processes. In order to fulfil its objectives, archaeology seeks the help of many other disciplines: anthropology, history, geology, physics, chemistry, biology, etc. For the present publication the past is defined in a very broad way, be it fifty or a thousand years ago. Archaeology discerns many different periods, based on geographical, chronological, cultural or evolutionary parametres. There may be South American, prehistoric, Roman or Iron Age archaeology, each of them—and many more—with its own specifics.

They may be important at the time of actually conducting archaeological research, but not for this kind of general discussion. Age is not necessarily a factor in determining the importance of archaeological remains. The standing remains of the past may have a fair chance of surviving, at least partially, if the population is respectful of them, through identifying themselves with the ruins of their ancestors. Unfortunately, architectural remains often serve as quarries for cheap building materials. The finest Inca masonry embellishes modern chimneys in Peru, and Indian carpentry becomes incorporated as decorative elements of houses in the west. These are just two simple examples of a much wider problem. In a way, it may be a minor problem, because of the size, weight or visibility of the objects being moved. On the other hand, there are hundreds of thousands of less spectacular, though even more destructive, pillages going on. Most of these concentrate on the buried remains of the past, with a marked preference for ancient graves. Millions of funerary objects can be found on the shelves of museums and private collections, be they ceramic pots, textiles, basketry, metals or any other artefact that accompanied the dead in the afterlife.

The great majority is the product of illegal excavations and trade between developing countries and the rich nations. On both sides, unscrupulous people have established the necessary channels to plunder the archaeological heritage for the sole sake of economic profit, thereby destroying beyond repair what may be unique contexts. Although archaeological cemetreies may be particularly vulnerable to this greed, a lot of other situations are disastrous for the buried archaeological patrimony. Most big building projects, such as roads and dams, don't take into account the existence of archaeological sites, and numerous controversies have arisen where ancient remains have been detected in time, holding up progress and causing huge economic losses.

But a lot of damage can also occur on the level of the individual citizen. It covers the wide range from innocent sherd-picking on archaeological sites to organised treasure hunting, from souvenir buying to illegal traffic in antiquities. All these destructions seriously limit the potential for discovering lost knowledge that may be of use today. Quantifying the amount of trade in cultural heritage will always be difficult, since an important part of it occurs clandestinely. However, recent estimates speak of US$ 1 billion worth of commerce, a great part of it made up by archaeological patrimony. This makes it second only to drug trafficking on world level. Some figures may illustrate the seriousness of the situation:

- In 1973 some 1 per cent of the economically active population of Costa Rica was estimated to be involved in illegal excavation or commerce of the results of it, representing about US$ 0.5 million annually (Walden, 1991).
- Of all thefts of cultural property, and in the case of illegal excavation probably less, only 12 per cent is ever recovered, meaning that over 88 per cent enters the illegal trade network (ibid.).
- One of the incentives to loot and sell the products can be found in the dazzling increase in prices for art and cultural property. The most important auction houses report incredible growths in revenue over the last ten years. Sothebys increased 800 per cent, Christies 700 per cent.

Archaeological Heritage

We look at the archaeological heritage as an important resource for sustainable development, a resource that is both instructive and finite, and thus irreplaceable. We therefore urgently need to consider the cultural aspect of our environment as essential as the ecological one. We must work out ways to protect the cultural environment from becoming as deteriorated as the natural environment now is. We suggest that one way of doing this is to define our environment in an integrated way, meaning that we are talking of a cultural-ecological environment. A few general reasons underlie this thinking:

- The natural environment doesn't have a lot of meaning if it doesn't include all of its components, a key player being man. The mere fact that the human element is now jeopardising its natural environment, makes it all the more interesting to investigate how such a situation was arrived at and, more importantly, how it used to be, when man and nature had a more symbiotic way of life. Maybe history can show us similar problems in the past and teach us how they were solved.

- Ecological preservation has hit the news for decades now, but it used to be completely different. We remember the times when almost anything was allowed in order to attain a higher level of development, including destroying the natural environment, resulting in the situation we have now. So let us not commit the same error on the cultural level, as we need it for assuring the right quality of life of future generations.
- There is a perceived need for safeguarding biological diversity, the rich natural variety of living species, which should be transposed to the cultural level. The many different cultural expressions represent the variety of adaptive responses to equally different situations, both natural and social, that can, ultimately, teach us solutions to present-day problems. So preserving the remains that illustrate the cultural diversity can have modern applications.
- Cultural identity is a fundamental need of populations that are undergoing major processes of change, as is the case of developing countries trying to catch up. Their feeling of proper identity must be strong enough to assimilate foreign inputs and to retain their own, while being constantly influenced and measured by international, universal standards that have no meaning whatsoever in their native world-view. Historical background, reflected by the archaeological remains, represents a key factor in cultural identity.

These reasons notwithstanding, modern-day measures for protecting the archaeological heritage leave much to be desired. Although legislation exists in almost every country, enforcement is usually very low key on both sides of development. If Italy still struggles with its tombaroli grave looters, one can imagine what the situation is in developing countries. Alarming, and often very sad, reports come from everywhere: primarily Middle and South America and Africa, but also South-East Asia. In all these regions, countries are facing the ever-increasing pillaging of their archaeological sites. Manifestly the most efficient protection measure is expropriation of the archaeological site, turning it into an archaeological park with appropriate infrastructure for tourism or recreation. But only very few sites qualify for this kind of treatment, because of the high costs involved and the often very complicated legal aspects.

Levels of Decision-making

Once the importance of the cultural heritage in general, and the archaeological patrimony in particular, has been accepted, one must ensure that it will be taken into account when designing development plans. First of all, of course, the political will must exist to redefine the environment as a dual concept that comprises both natural and cultural aspects. This battle against traditional theories and conservative bureaucracies may not be easy, but the concept of sustainability in all its facets should be of paramount usefulness. Its emphasis on social and economic aspects, besides the ecological one, should allow for broadening the scope of the definition and, as a corollary, the scope of action.

We must beware of being overoptimistic: the systematic inclusion of cultural aspects, let alone the archaeological heritage, in the broad field of development planning will indeed be a very hard task. A comparison with ecological conservation can be instructive. Even now, after decades of warnings, the systematic screening of development projects on their environmental consequences is only slowly finding its way into the higher levels of decision-making. The discussion about extending the environmental impact assessment (EIA), on the project level, to the strategic environmental assessment (SEA) on the policy level, has only just been initiated and promises to be lengthy. We can only hope that the new ground being broken there will provide a path for our purposes.

At the highest level then, policy-makers need to be convinced, but regional authorities and local people are equally important. Indeed, many times already the initiatives that eventually lead towards the execution of development projects have originated within the population that, later, will become the target group of those projects. Development should always happen this way. Problems should be recognised by those who suffer their consequences, and be communicated to the next level of hierarchy, which should take action to solve the problems. Too many times in the past, and even now, existing problems were perceived at a high level of decision-makers who then turned to external sources for help, both economic and technical, without a thorough knowledge of all the aspects of the problem in question. As a result the solution offered often failed to meet the expectations and the real needs of the population involved.

The input of the local population is especially indispensable when considering cultural aspects and archaeological heritage. High-level policy-makers may belong to an entirely different cultural group, or may simply ignore the existence of archaeological remains. Many reasons may exist for not taking these elements into account, most commonly lack of interest or fear of complicating the action under consideration. This does not mean that the central authority cannot have good ideas for developing local groups; it does mean that close co-operation between the local population and the authorities at all levels, from the very beginning of development planning, will undoubtedly lead to easier execution and positive results. External donors should be very much aware of this mechanism and, without intending to dictate to the developing countries, should definitely try to promote local involvement in decision-making.

When speaking of local authorities, a wider range than the official authorities should be envisaged. Sometimes, government officials at the local level are people brought in from outside the region and may lack sympathy with the local group and its historical roots. In parallel, there may be an ethnic local government, according to traditional social organisation. It is not uncommon to have several tribal chiefs, ethnic lords, or whatever, living under the authority of a representative from the central government. These chiefs themselves head a complex social structure, not necessarily compatible with the official government structure, and are probably acting as buffers between the official and the ethnic sides. When it comes to implementing development plans successfully, the local ethnic players may prove to be of decisive importance. Their familiarity and identification with the region, both ecologically and historically, their knowledge of ancient technology and archaeological sites will all be assets in designing and implementing integrated development plans.

The intervention Cycle

If we are serious about giving consideration to the cultural aspects of a society, and among them the historical testimonies of its identity, then we need to do so from the outset. Any intervention in a country, region or human group should evaluate the possible consequences for the archaeological heritage of the involving people target population, in exactly the same way that the impact on the natural environment is being assessed.

The community of Agua Blanca, located in the valley of Rio Buena Vista, lies in the southern part of the Province of Manabi, on the coast of Ecuador. This small community, composed of some 40-odd families, living in a small hamlet and some scattered houses in the valley, subsisted traditionally on a mixed economy of fishing, some cattle-herding and limited agriculture.

Being situated in the heartland of the pre-Columbian Manteño culture (AD 500-1500), surrounded by a fair quantity of architectural remains, an important complementary income was generated by pot-hunting in the archaeological sites.

When the area was incorporated into the Machalilla National Park in 1979 and serious restrictions were imposed on the exploitation of natural resources, archaeological looting practically became the main revenue for the local population, in spite of the fact that one of the park's objectives was to preserve the archaeological resources. Regularly, recognisable Manteno objects from the region invaded the antiques market, a practice that continued for several years.

It wasn't until the 1980s, when a British archaeologist became interested in the area and started designing a research project that the local people became aware of the damage they were inflicting on their heritage. By explaining the importance of the ancient remains, showing their extension in the valley and beyond, and inferring the economic and social situation, he was able to raise the locals' interest in their 'ancestors'.

When starting his research, the archaeologist hired the local people to participate in the survey and excavations. During fieldwork frequent comments were heard about similar finds the locals had made during their looting campaigns and the archaeologist could gather a lot of additional information. He was even allowed to see several personal collections and learned about others.

Over time, becoming ever more acquainted with the local population and its problems, the archaeologist managed to finance certain improvements in the infrastructure of the valley and took effective steps to promote the archaeological heritage, so that tourism started to increase.

After the fieldwork and the processing of the archaeological finds, the local people became involved in the maintenance of the site and finally ended up attending the visitors. In order to present a better view of the history and the archaeological heritage of the region, a museum was built and was equipped with a few objects stemming from the excavations and…a lot of artefacts donated by the local people. Nowadays, the daily care of site and museum is in the hands of the community. The income from entrance fees and from the selling of handicrafts has completely replaced that from archaeological looting. Additionally, the Agua Blanca community is known as an important defender of the archaeological heritage, organising meetings with other communities in order to raise awareness through communicating their own experience.

This means that an archaeological impact assessment should be considered, as an integral part of EIA. Ideally, this exercise should begin as early as the identification phase of an intervention, but should continue throughout the rest of the cycle. Indeed, not everything can be foreseen when it comes to evaluating the archaeological heritage of an impact area, given that many elements may be buried, and trying to detect the totality of them beforehand may be beyond the overall scope of the intervention. Flexibility during the whole project cycle thus becomes a very important characteristic. Flexibility in general should always be a prime concern when designing and executing development plans.

The screening of interventions for their eventual archaeological impact should be integrated into the systematic environmental screening, by including specific questions that seek general information on visible or known remains from the past. Further key questions relate to the amount of subsurface intervention that is planned, as it may affect buried elements. The quantity and quality of the remains must then be evaluated and safeguarding priorities established. Not everything can be preserved but a representative sample should be protected by changing the intervention modalities or localities. A few exceptional sites may be worth saving from destructive large-scale projects in the way UNESCO did at the time of the building of the Aswan dam.

If, at an early stage, the projected activities of the intervention suggest that part of the archaeological heritage will be affected in a negative way, specific actions need to be taken in order to avoid or minimise impact. If no danger exists, the intervention can go forward. If there are uncertainties about the type and extent of the archaeological heritage that will be affected, then further study is warranted. Ultimately, an entire project could be cancelled because of unjustifiable destruction. Before such a harsh decision is taken, however, alternatives to the original project should be sought or redesigning of the project should be intended.

Impact Assessment

As soon as the archaeological elements of a region are considered to be an integral part of the environment, then any AIA should not be separated from the general EIA. It would just be another dimension of EIA, to be handled by an archaeologist within the EIA team, since only a professional can have the accurate knowledge of both the existing information and the techniques actively to discover ancient remains.

The first step is getting acquainted with the intervention area and the objectives of the project. This should give initial indications about a potential impact on the archaeological heritage. Many interventions may be rapidly qualified as non-destructive, so that no further action is required. Projects in sectors such as public health and education rarely jeopardise archaeological remains. In general, only interventions that modify the landscape, including any form of digging, should be closely watched.

During the second step, unless the archaeologist in charge is completely familiar with the region, time will be needed to study the specialised literature. Information may be found in old maps, travel accounts, early geographical descriptions, historical archives, former official documents, excavation reports, anthropological studies, etc. All elements thus discovered should be mapped provisionally and their existence and state of conservation verified in the field. Interviewing people who know the area and its history well may increase the preliminary inventory of archaeological remains. At this point, a better idea of the potential impact can emerge. Once the verification and final mapping is completed, certain patterns in the ancient remains, or the lack of them, may show up and a decision on the need for further research must be taken.

Up to this point the assessment may be considered as a standard screening and scoping procedure, identical to environmental screening and resulting in a broad classification of the kind 'no impact', 'low impact', 'high impact'. In addition, however, there may often be insufficient information available. Indeed, lack of information in archaeology doesn't automatically mean that there are no remains; usually it means that they haven't been found yet, mostly because they are buried and thus invisible to the naked eye. The decision on whether or not to continue investigating will be difficult, since any further research will be more specialised, longer term and more expensive.

The next step represents the active search for archaeological remains through a wide variety of surveying techniques. Where available and applicable, remote sensing is a great tool for having easy access to extended areas. Aerial photography, originally used for national cartographic purposes, is an important source for detecting monumental remains. Satellite images can do the same with an ever-increasing degree of detail and resolution. Low altitude aerial photography can even show buried remains through soil marks, crop marks or shadow marks. But no survey would be complete or accurate without a pedestrian phase, walking the area while scrutinising the surface for remains of buried sites. In order to limit time and expenses, several sampling techniques can be applied. If a realistic image of the archaeological heritage is being pursued, then test excavations must be conducted.

Only in this way can the depth and the characteristics of archaeological components of the region be established. Now the archaeological picture should be sufficiently clear to evaluate the real impact of the planned intervention. The archaeological impact assessment in itself can be considered to be finished, but mitigating actions may be warranted. Setting aside an eventual follow-up, it should be clear that, at this stage, already a lot of energy and resources will have been allocated while the possibility of cancelling the intervention still lures. This may result in very tough decisions to take, much in the same way as applies to the ecological environment, but sooner or later this is the point we should reach. If planning starts a long time ahead and if it is integral, that is integrating all aspects of sustainable development, then a lot of surprises can be avoided and enough time will remain to either conduct the necessary research or look for alternatives. Ultimately, the real anticipatory approach, as stipulated by SEA, should prevail, thereby transcending the narrow and restrictive project-by-project approach.

Once the archaeological picture is crystal clear, decisions with regard to the project must be taken. Several options may be present and we shall run through them briefly, from no special action required to project cancelled. After an exhaustive study of the project impact area there may be no evidence at all of standing or buried archaeological remains. Obviously no special action is then required and the planned project can go ahead unmodified. The archaeological study may have come across remains of the past that will be impacted by the project activities as originally set out. Different possibilities for the conservation of the cultural elements arise:

1. They may be left untouched by changing the project.
2. They may be safeguarded against the effects of the project.
3. They may be saved by removing them from their original location.
4. They may be destroyed.

In order to arrive at the best solution, many factors must be studied, compared and weighed for each option and a lot will depend on the kind of project activity and the type of archaeological remain. In order to illustrate these options let us suppose that a development project will inundate part of a valley with the aim of providing drinking water, irrigation and electric power to the populations downstream. After having studied the area archaeologically, the project may be approved on the condition that certain modifications be introduced and complementary activities be taken up. The modified version of the project may reduce the height of the dam, limiting the flooding so that archaeological sites may be left untouched (1). The most important remains may be surrounded by a protective wall, creating sunken islands (2), or they may be dismantled, transported to higher grounds and reassembled (3). A few sites, considered less important because they are repetitious or have been studied thoroughly, could be sacrificed (4).

However, a site should never be destroyed without being properly registered, meaning that its location and characteristics are carefully described.

Because each archaeological site has a story to be told, destruction should be the last resort, conservation being the prime target. Nevertheless, trying to preserve every single testimony of the past is utopian and undesirable. It would turn the whole world into a showcase, denying the right to progress. The ultimate goal of conservation is sustainable development, not stagnation by exaltation of antiquity.

It should be clear by now that any form of AIA and its follow-up will inevitably cause additional time and money requirements on the part of the project. These incremental costs should be considered in the same way as the present environmental impact consequences of development projects. Funding facilities should be available for cultural/archaeological impact mitigation, much in the manner they operate for the impact on the ecological environment.

In this sense it is a promising perspective that the World Bank encourages borrowers to incorporate into their environmental action plans any major cross- sectoral issue, including cultural and natural heritage. The World Bank considers the conservation of important cultural property to be part of the sustainable development process because the important cultural values of a society or social group are frequently reflected in cultural property.

Also UNDP has expressed and practised the linkage between its global urban development co-operation strategy and the cultural preservation of historic cities, especially in Latin America and the Caribbean.

Protecting Sites for Conservation

Given the importance of the archaeological heritage as a key factor in the identity of people and as a potentially rich source for traditional knowledge and know-how, given also the possible omnipresence of the archaeological patrimony and therefore its proneness to destruction, given finally the ever- increasing lack of already scarce resources for research and ad hoc conservation, mechanisms should be developed to go beyond the project-oriented approach, that is to anticipate possible problems during project preparation, much in the way that the EIA approach is now trying to move towards an SEA approach.

The strategic approach in archaeological conservation would involve vast programmes of archaeological reconnaissance and survey, in order to establish the extension and characteristics of the heritage, followed by effective and efficient measures to assure its long-term conservation. Indeed, if we want to protect our common patrimony, we first need to know precisely where it is located and what the most immediate dangers are.

We therefore suggest we start out around big cities or other places that are expanding rapidly so that once specific urbanisation plans appear we already know how to adapt them to the situation of the archaeological heritage.

Important archaeological sites or areas may eventually become real archaeological parks that are well-protected areas where any kind of disturbing activity is banned or controlled by law. The number of these parks, at any time, will probably be limited and reserved for very special monuments, but a fair amount of archaeological sites may be combined with existing ecological reserves or natural parks, thereby enhancing these with the human aspect, while profiting from the protective ruling. Where the previous solutions are not feasible, archaeological sites should be turned into green areas within the urban perimetre. The kind of activities allowed in these places will depend on the characteristics of the site. Standing remains should be protected against direct contact. Buried sites could serve as a recreation area as long as interference with the subsurface archaeological layers is avoided. Sometimes adding a protective cover of topsoil may create a perfect archaeological reserve, while allowing some social function.

Monitoring

Once a clear picture of the archaeological heritage exists and the different degrees of conservation categories, with their specific conservation measures, have been established, a matching system of monitoring should be devised. Monitoring is here understood to be the periodic evaluation of the state of conservation of an archaeological site.

According to the kind of site and the protection measures that apply to it, monitoring can range from a simple inspection visit in the case of public green areas, to a specialised and integral monitoring mission when checking on important monuments. It should be carried out by a specialised national institution, normally the same one that establishes and assigns the conservation categories.

In any case, special care must be given to an adequate level of reporting of the monitoring exercise, developing a standardised reporting system so that the information gathered over time may be comparable. That information should include aspects other than the mere physical condition of the site, covering its use and abuse, its social and economic functions or potentials, paying special attention to the informal economy that tends to build up around tourist places.

Quito, the capital of the Republic of Ecuador, is located on a step-like terrace of the interandean valley, that runs North-South along the Western Cordillera. A long-stretched city of a little over 30 km long by 5 or 6 km wide, at 2,800 m above sea level, and almost exactly on the Equator, it is endowed with year-round spring weather and a rich variety of ecological niches in the immediate surroundings.

No wonder human occupation in and around Quito is among the oldest in Ecuador, dating back as far as 7000 BC with well-established settlements in what is now the city from 1500 BC onwards. The Neo-lithic village of some 26 hectares and 700 inhabitants has now become a capital that occupies approximately 20,000 hectares and houses 1.1 million people, and is growing rapidly. The Municipal Planning Unit has analysed the historical information on urban growth over the last 85 years, establishing a decreasing growing rate, but still projecting a doubling of the urban area to over 40,000 hectares by 2020, and a tripling of the population to almost 3.5 million. On the other hand, recent archaeological research in the Quito Valley by the Belgian Agency for Development Co-operation has established a density of one archaeological site per 2 km2, based on a systematic regional survey by sampling.

The conclusion from both types of information can only be that over the next 20 years a total of 100 archaeological sites will be lost, due exclusively to the urban expansion of the capital. Therefore, a project was suggested, and will hopefully be implemented some day, that would develop the capacity of the municipal services to anticipate the emergency situations that will occur once effective urbanisation starts.

The project would basically aim at the organisation of a municipal archaeological service which, in close co-ordination with the office for urban planning, would design a survey programme systematically to cover all the areas that will eventually become incorporated into the city. The presence and characteristics of ancient sites would be registered and evaluated in order to assign conservation categories and priority degrees for their protection. According to this system, urbanisation plans will have to take into account the archaeological heritage and adapt to it. The project would also contemplate campaigns of awareness-raising in the other municipal services and at the level of the general public.

Although the basic research preceding this conservation classification, i.e. survey and excavations, will require specific human and financial resources, positive outputs may be expected in terms of understanding the region's archaeology, and in terms of significant economic gains at the time of urbanisation, by avoiding costly delays in work already started.

Awareness-raising and Capacity-building

In a field that is as new as environmental management and archaeology, much awareness-raising and capacity-building is required. In many developing countries a lot has still to be done on the level of basic archaeology teaching. Indeed, with a few notorious exceptions, professional training leaves much to be desired and when formal teaching exists, it is often concentrated on the theoretical level, leaving the students with hardly any field experience by the time they graduate. Whenever these young professionals get involved in emergency situations or regular research, their work clearly shows those training deficiencies.

This situation has originated a kind of intellectual or cultural colonialism on the part of foreign scholars or graduate students, who are eager to find new fields of study in the developing countries. They used to invest time and money to explore these lesser-known regions and become experts and founders of new archaeology departments at home. Fortunately, many of the early explorers have taught local people in the field or have provided grants to study abroad. However, even today we must ascertain that there still are, mostly at the doctoral dissertation level, European and American graduates who exploit intellectually the archaeological heritage of a developing country by conducting their research without feeding back its results to the country that generously hosted them.

In order to take up the archaeological heritage in the environmental management, capacity-building should be applied to different levels. At the highest level, political will must be created by informing the policy-makers about the importance of the cultural and archaeological heritage in identity-building and sustainable development, emphasising its educational role and taking into account its economic potential. Ministries of culture and education, national institutes for the cultural heritage or national museums for history and archaeology should take the lead in these awareness-building, seeking examples of proven good practices in other countries. Enhancing the capacity to cope with the more practical action plans on the national level may require specific assistance from outside.

In the professional realm, a quality basic training in archaeology at the national university level should be assured, complemented with specialisation possibilities abroad, through grant agreements with foreign academic or technical institutions. Emphasis should be put on the interdisciplinary approach of archaeological research and on the social function the discipline should pursue, beyond the pure scientific immediate results. General notions about the objectives and functions of archaeology should also be introduced in related disciplines, especially those that are frequently in the field that is natural sciences but also, and very importantly, urban planning and general development.

At the service level, much archaeological information is lost through lack of reporting or destruction by ignorance, due to the fact that fieldworkers of companies or institutions that take care of electricity and telephone distribution, of water or sewer infrastructure, etc. have not the slightest clue to the importance of the remains they come across. These people especially are in constant contact with subsurface layers in the urban and rural area, and countless stories have been heard of the many fortuitous finds they have discovered during work. Workers and field supervisors, as well as the higher administrative levels of these companies and institutions should therefore receive some basic knowledge of how the past may appear in their daily work, who they should report to and what emergency measures eventually could be taken.

Finally, at the level of general education, starting in primary school and from there up, the young should not only learn about their country's history on a textbook basis, but should be confronted directly with whatever remains of that history are left. Raising awareness about the importance of conserving the evidence of past cultures should be developed together with building up the ecological conscience. Basic education probably represents the most important level on which to work for this purpose, since it means both investing in the future—this youth will soon become the main actor in sustainable development—and creating a multiplicator effect through the influence youngsters exert on their parents.

International Co-operation

Our self-imposed programme for the future, Agenda 21, covers a wide array of hopes and aspirations for a better future, intrinsically relating the environment to the economics and social aspects of development. Several legally binding offsprings of the Rio Conference are well underway in taking effective and efficient measures to put ideas into action. However promising and necessary all this is, the cultural dimension of development is still missing.

Little regard is given to the uniqueness of the social systems that managed the environment in ways that were much more sustainable than our 'modern' alternatives. At the most, brief references are made to the need for recuperating local know-how, but only very rarely is mention made of the preservation of indigenous culture. Since no precise definitions are usually given, the question of how culture in this context is exactly conceived remains wide open. One strongly senses that it is rather limited to folk culture, the present 'exotic' aspects of foreign people, not the rich totality of traditional beliefs, knowledge, art, technology, etc. that has its most genuine expression in the archaeological heritage.

One of the reasons for this lack of cultural dimension may be due to the fact that other bodies have been involved in these problems at an earlier stage and, at the same time, from a more technical point of view. The international organisation that first comes to mind is UNESCO, which adopted the World Heritage Convention in 1972—the same year of the Stockholm Conference on the Environment—in order to protect the natural and cultural heritage against increasing threats of destruction by natural and human causes. Earlier conventions deal with the protection of cultural heritage in armed conflict and with the measures to prevent illegal import, export and transfer of cultural heritage. So, apparently, several international agreements are taking care of the cultural and archaeological heritage. In effect, a lot has been done, but past and present efforts were not able to, and still can't, keep up with the ever-increasing pace of development and the subsequent destruction of ancient remains. UNESCO alone can't handle the situation, although it has been building partnerships on a regional or national basis everywhere.

Environment Policy and Environment Management System

Driven by its commitment for sustainable growth of power, NTPC has evolved a well defined environment management policy and sound environment practices for minimising environmental impact arising out of setting up of power plants and preserving the natural ecology.

National Environment Policy

At the national level, the Ministry of Environment and Forests had prepared a draft Environment Policy (NEP) and the Ministry of Power along with NTPC actively participated in the deliberations of the draft NEP. The NEP 2006 has since been approved by the Union Cabinet in May 2006.

NTPC Environment Policy

As early as in November 1995, NTPC brought out a comprehensive document entitled "NTPC Environment Policy and Environment Management System". Amongst the guiding principles adopted in the document are company's proactive approach to environment, optimum utilisation of equipment, adoption of latest technologies and continual environment improvement. The policy also envisages efficient utilisation of resources, thereby minimising waste, maximising ash utilisation and providing green belt all around the plant for maintaining ecological balance.

Environment Management, Occupational Health and Safety Systems

NTPC has actively gone for adoption of best international practices on environment, occupational health and safety areas. The organization has pursued the Environmental Management System (EMS) ISO 14001 and the Occupational Health and Safety Assessment System OHSAS 18001 at its different establishments. As a result of pursuing these practices, all NTPC power stations have been certified for ISO 14001 and OHSAS 18001 by reputed national and international Certifying Agencies.

Pollution Control Systems

While deciding the appropriate technology for its projects, NTPC integrates many environmental provisions into the plant design. In order to ensure that NTPC comply with all the stipulated environment norms, various state-of-the-art pollution control systems/devices as discussed below have been installed to control air and water pollution.

Electrostatic Precipitators

The ash left behind after combustion of coal is arrested in high efficiency Electrostatic Precipitators (ESPs) and particulate emission is controlled well within the stipulated norms. The ash collected in the ESPs is disposed to Ash Ponds in slurry form.

Flue Gas Stacks

Tall Flue Gas Stacks have been provided for wide dispersion of the gaseous emissions (SOX, NOX etc) into the atmosphere.

Low-NOX Burners

In gas based NTPC power stations, NOx emissions are controlled by provision of Low-NOx Burners (dry or wet type) and in coal fired stations, by adopting best combustion practices.

Neutralisation Pits

Neutralisation pits have been provided in the Water Treatment Plant (WTP) for pH correction of the effluents before discharge into Effluent Treatment Plant (ETP) for further treatment and use.

Coal Settling Pits/Oil Settling Pits

In these Pits, coal dust and oil are removed from the effluents emanating from the Coal Handling Plant (CHP), coal yard and Fuel Oil Handling areas before discharge into ETP.

DE and DS Systems

Dust Extraction (DE) and Dust Suppression (DS) systems have been installed in all coal fired power stations in NTPC to contain and extract the fugitive dust released in the Coal Handling Plant (CHP).

Cooling Towers

Cooling Towers have been provided for cooling the hot Condenser cooling water in closed cycle Condenser Cooling Water (CCW) Systems. This helps in reduction in thermal pollution and conservation of fresh water.

ASH DYKES AND ASH DISPOSAL SYSTEMS

Ash ponds have been provided at all coal based stations except Dadri where Dry Ash Disposal System has been provided. Ash Ponds have been divided into lagoons and provided with garlanding arrangements for change over of the ash slurry feed points for even filling of the pond and for effective settlement of the ash particles. Ash in slurry form is discharged into the lagoons where ash particles get settled from the slurry and clear effluent water is discharged from the ash pond.

The discharged effluents conform to standards specified by CPCB and the same is regularly monitored. At its Dadri Power Station, NTPC has set up a unique system for dry ash collection and disposal facility with Ash Mound formation. This has been envisaged for the first time in Asia which has resulted in progressive development of green belt besides far less requirement of land and less water requirement as compared to the wet ash disposal system.

ASH WATER RECYCLING SYSTEM

Further, in a number of NTPC stations, as a proactive measure, Ash Water Recycling System (AWRS) has been provided. In the AWRS, the effluent from ash pond is circulated back to the station for further ash sluicing to the ash pond. This helps in savings of fresh water requirements for transportation of ash from the plant.

The ash water recycling system has already been installed and is in operation at Ramagundam, Simhadri, Rihand, Talcher Kaniha, Talcher Thermal, Kahalgaon, Korba and Vindhyachal. The scheme has helped stations to save huge quantity of fresh water required as make-up water for disposal of ash.

Dry Ash Extraction System (DAES)

Dry ash has much higher utilization potential in ash based products (such as bricks, aerated autoclaved concrete blocks, concrete, Portland pozzolana cement, etc.). DAES has been installed at Unchahar, Dadri, Simhadri, Ramagundam, Singrauli, Kahalgaon, Farakka, Talcher Thermal, Korba, Vindhyachal, Talcher Kaniha and BTPS.

Liquid Waste Treatment Plants and Management System

The objective of industrial liquid effluent treatment plant (ETP) is to discharge lesser and cleaner effluent from the power plants to meet environmental regulations. After primary treatment at the source of their generation, the effluents are sent to the ETP for further treatment. The composite liquid effluent treatment plant has been designed to treat all liquid effluents which originate within the power station e.g. Water Treatment Plant (WTP), Condensate Polishing Unit (CPU) effluent, Coal Handling Plant (CHP) effluent, floor washings, service water drains etc.

The scheme involves collection of various effluents and their appropriate treatment centrally and re-circulation of the treated effluent for various plant uses. NTPC has implemented such systems in a number of its power stations such as Ramagundam, Simhadri, Kayamkulam, Singrauli, Rihand, Vindhyachal, Korba, Jhanor Gandhar, Faridabad, Farakka, Kahalgaon and Talcher Kaniha. These plants have helped to control quality and quantity of the effluents discharged from the stations.

Sewage Treatment Plants and Facilities

Sewage Treatment Plants (STPs) sewage treatment facilities have been provided at all NTPC stations to take care of Sewage Effluent from Plant and township areas. In a number of NTPC projects modern type STPs with Clarifloculators, Mechanical Agitators, sludge drying beds, Gas Collection Chambers etc have been provided to improve the effluent quality.

The effluent quality is monitored regularly and treated effluent conforming to the prescribed limit is discharged from the station. At several stations, treated effluents of STPs are being used for horticulture purpose.

ENVIRONMENTAL INSTITUTIONAL SET-UP

Realizing the importance of protection of the environment with speedy development of the power sector, the company has constituted different groups at project, regional and Corporate Centre level to carry out specific environment related functions.

The Environment Management Group, Ash Utilisation Group and Centre for Power Efficiency and Environment Protection (CENPEEP) function from the Corporate Centre and initiate measures to mitigate the impact of power project implementation on the environment and preserve ecology in the vicinity of the projects. Environment Management and Ash Utilisation Groups established at each station, look after various environmental issues of the individual station.

Environment Management During Operation Phase

NTPC's environment friendly approach to power has already begun to show results in conservation of natural resources such as water and fuel (coal, oil and gas) as well as control of environmental pollution. As already mentioned earlier, NTPC has chalked out a set of well defined activities that are envisaged right from the project conceptualisation stage so that during the entire life cycle of the power plant, NTPC is fully compliant with various environment regulations and a pristine environment and ecological balance is maintained in and around its power station and townships.

Following is brief description of some of the measures taken during the operation phase of the stations. Performance enhancement and up-gradation measures are undertaken by the organisation during the post operational stage of the stations. These activities have greatly helped to minimise the impact on environment and preserve the ecology in and around its power projects. These measures have been enumerated as follows.

Monitoring of Environmental Parameters

A broad based Environment Monitoring Programme has been formulated and implemented in NTPC. All pollutants discharged from the power plant such as stack emission, ash pond effluent, main plant effluent, domestic effluent and Condenser Cooling Water (CCW) effluent are monitored at the stipulated frequency at the source itself and at the points of discharge.

In addition to the above, ambient air, surface water and ground water quality in and around NTPC plants are regularly monitored to assess any adverse impacts as a result of operation of the power plant.

On-Line Data Base Management

In order to have better control on pollution and to achieve effective environment management in and around NTPC stations, it is imperative to have an on-line, reliable and efficient environment information system on the plant operational and environmental performance parameters at all three levels i.e generating Stations, Regional Headquarters and Corporate Centre. In consideration of the above, a computerized programme, namely "Paryavaran Monitoring System" - PMS, which could provide reliable storage, prompt and accurate flow of information on environmental performance of Stations was developed and installed in NTPC.

This software facilitates direct transfer of environment reports and other environment related information from stations to the Regional Headquarters and Corporate Centre.

The PMS has already been implemented at Corporate Centre, the Regional Headquarters and most of the Stations. This system has helped in achieving continuous improvement in NTPCs environment performance through improved monitoring and reporting system by using the trend analysis and advanced data management techniques.

ENVIRONMENT REVIEWS

To maintain constant vigil on environmental compliance, Environmental Reviews are carried out at all operating stations and remedial measures have been taken wherever necessary.

As a feedback and follow-up of these Environmental Reviews, a number of retrofit and up-gradation measures have been undertaken at different stations. Such periodic Environmental Reviews and extensive monitoring of the facilities carried out at all stations have helped in compliance with the environmental norms and timely renewal of the Air and Water Consents.

Upgradation and Retrofitting of Pollution Control Systems

In order to keep pace with the changing norms and ensure compliance with statutory requirements in the field of pollution control, NTPC keeps an open mind for Renovation and Modernisation (R and M) and Retrofitting and Upgradation of pollution monitoring and control facilities in its existing stations.

It is important to mention that such modifications/retrofit programmes not only helped in betterment of environment but also in resource conservation. High efficiency Electro-Static Precipitators (ESPs) of the order of 99.5 per cent and above have been provided at NTPC stations for control of stack particulate emissions.

However, the ESPs of a number of stations were built prior to the promulgation of the Environment (Protection) Act, 1986 and notification of emission control standards under this Act. Remedial measures have already been taken up and implemented to improve the efficiency of the existing ESPs at various NTPC stations. ESP performance enhancement programme by adopting advanced microprocessor based Electrostatic Precipitator Management System (EPMS) was installed at its power stations at Singrauli, Ramagundam, Korba, Farakka, Rihand, Vindhyachal and Unchahar. Additional ESPs were retrofitted in the older power stations, namely at Badarpur and Talcher Thermal.

As a result of the above retrofits, the emission of Suspended Particulate Matter (SPM) has been brought down appreciably at the above stations and is maintained within the present statutory limit of 150 mg/Nm3. In new projects, the ESPs have been designed for a maximum permissible outlet dust emission of 50 mg/Nm3 to meet the likely stringent emission norms in the near future.

RESOURCES CONSERVATION

With better awareness and appreciation towards ecology and environment, the organization is continually looking for innovative and cost effective solutions to conserve natural resources and reduce wastes.

Some of the measures include:

- Reduction in land requirements for main plant and ash disposal areas in newer units.
- Capacity addition in old plants, within existing land.
- Reduction in water requirement for main plant and ash disposal areas through recycle and reuse of water.
- Efficient use of Fuel (Coal, Natural gas and Fuel oil)
- Reduction in fuel requirement through more efficient combustion and adoption of state-of-the-art technologies such as super critical boilers

WASTE MANAGEMENT

Various types of wastes such as Municilal or domestic wastes, hazardous wastes, Bio-Medical wastes get generated in power plant areas, plant hospital and the townships of projects. The wastes generated are a number of solid and hazardous wastes like used oils and waste oils, grease, lead acid batteries, other lead bearing wastes (such as garkets etc.), oil and clarifier sludge, used resin, used photochemicals, asbestos packing, e-waste, metal scrap, C&I wastes, electricial scrap, empty cylinders (refillable), paper, rubber products, canteen (bio-degradable) wastes, buidling material wastes, silica gel, glass wool, fused lamps and tubes, fire resistant fluids etc.

These wastes fall either under hazardous wastes category or non-hazardous wastes category as per classification given in Government of Indias notification on Hazardous Wastes (Management and Handling) Rules 1989 (as amended on 06.01.2000 and 20.05.2003). Handling and manegement of these wastes in NTPC stations have been discussed below.

Municipal Waste Management

Domestic or municipal waste is generated in households at townships. This waste is seggregated into bio-degradable and non-biodegradable wastes at source itself in different coloured containers and thereafter the two types are disposed separately. Bio-degradable waste is spread uniformly in identified low lying areas and thereafter it is covered with soil for use later as manure after composting. The seggregated non-bio-degradable waste is disposed off separately in other identified low lying areas and is spread out uniformly.

Hazardous Waste Management

NTPC being a proactive organization, the handling and disposal of hazardous wastes are done as per the Hazardous Wastes (Management and Handling) Rules 1989 (as amended in 2003) guidelines issued by Government of India for the treatment, storage and disposal of hazardous wastes. Scientific study on management and handling of hazardous wastes was carried out at a few NTPC stations to adopt the best practices so that there is a complete compliance with statutory requirements. In NTPC sataione, the Hazardous Wastes (Recyclable) are sold/auctioned to registered recyclers/refiners.

The other hazardous wastes such as the activated carbon resins, used drums (hazardous) chromium (Cr-III electrolytes, used petro-chemicals, asbestos packings, used torch batteries, ribbon, toners/cartridges, mixed wastes (waste oil, water and cotton) filters, earth contaminated with synthetic oil (FQF) glass used and sodium silicate, lamps and tubes etc. fall under the category of Hazardous Wastes (Non-Recyalable). These wastes are small in quantity and are stored in properly identified locations. As per the notification, hazardous wastes (non-recyalable) are to be sent to State Pollution Control Board (SPCB) approved common treatment storage and disposal facility (TSDF).

Bio-Medical Waste Management

Hospital (or Bio-medical) wastes get generated from hospitals and they include urine bags, human anatomical wastes, plaster of aris waste, empty plastic bottles of water and glucose, blood and chemical mixed cotton, blood and urines tubes etc. these wastes are segregated and are placed in buckets of different colours as per the notification for Bio-Medical Waste (Management and Handling) Rules.

The seggregated bio-medical wastes are either disposed through the SPCB approved agency or they are treated in autoclaves before disposal into bio- medical waste disposal pits. The treated bio-medical waste is spread uniformly and covered with 10 cm thick soil in bio-medical waste disposal pits.

LAND USE/BIO-DIVERSITY

As a policy, NTPC lays special emphasis on land use and Bio-diversity by way of development of green belts, energy plantations, reclmation of abandoned Ash Ponds and EIA and ecological monitoring in the project areas and its surroundings.

Reclamation of Abandoned Ash ponds

The reclamation of abandoned ash pond sites is a challenging task. NTPC has reclaimed temporary ash disposal areas at some of its projects namely Ramagundam, Talcher Thermal, Rihand, Singrauli and Unchahar through plantation and converted these sites into lush green environments. Extensive plantations have also been undertaken on dry ash mound at NTPC-Dadri. It is planned to reclaim all the abandoned ash disposal areas by plantation.

Green Belts, Afforestation and Energy Plantations

Whats more, in a concerted bid to counter the growing ecological threat, NTPC is undertaking afforestation programmes covering vast areas of land in and around its projects. Appropriate afforestation programmes for plant, township and green belt areas of the project have been implemented at all projects. In order to enhance green cover in the areas around our projects, as a responsible corporate citizen, NTPC till date has planted more than 18 million trees at its projects throughout the country.

The afforestation has not only contributed to the aesthetics but also has been serving as a 'sink' for the pollutants released from the station and thereby protecting the quality of ecology and environment in and around the projects. Thrust has also been given to bio-diesel plantation and around 4.8 lakh energy plants including Pongamia and Jatropha have already been planted. A pilot project for extraction of seeds from these bio-diesel plants has also been set up.

ECOLOGICAL MONITORING AND SCIENTIFIC STUDIES

NTPC has been a leader in the industrial sector of India in undertaking scientific studies related to thermal power generation. NTPC has pioneered several scientific studies in collaboration with national/ international institutions to develop an environmental databank e.g. Detailed Geohydrological Studies to understand the impact of ash pond leachate on ground water and Ecological Impacts Monitoring through Remote Sensing Data have been carried out at its operating stations as discussed below.

Environment Impact Asssement Studies

Environmental Impact Assessment (EIA) Studies are inevitably undertaken to evaluate potential negative impacts as well as to formulate Environmental Management Plans to overcome the identified impacts.

Based on the recommendations of Environmental Impact Assessment Study and Environmental Management Plan (EMP) and the conditions stipulated in the clearances from Ministry of Environment and Forests and State Pollution Control Boards, These studies consists of impact assessment in the area of the land use, water use, socio-economic aspects, soil, hydrology, water quality, meteorology, air quality, terrestrial and aquatic ecology and noise.

These studies are conducted before starting the construction as well as after operation of the plant and gives comprehensive status of the environment as existed before construction as well as in the post operational stages of the project. The EIA involves stage-by-stage evaluation of various parameters which affect the environment. Based on EIA study, wherever required, specific scientific studies are also conducted to scientifically assess the likely impact of the pollutants on the sensitive flora and fauna in the surroundings, as also, to take preventive and mitigatory measures, wherever required.

Socio-economic Studies

Detailed socio-economic studies are undertaken to establish the socio- economic status of project affected persons and rehabilitation and resettlement plans are drawn in consultation with the state government. Rehabilitation and resettlement options include land for land (subject to availability), limited jobs with NTPC and contractors and self employment schemes. In addition, NTPC also undertakes community development activities in the surrounding villages.

Ecological Monitoring Programme

NTPC has undertaken a comprehensive Ecological Monitoring Programme through Satellite Imagery Studies covering an area of about 25 Kms radius around some of its major plants. The studies have been conducted through National Remote Sensing Agency (NRSA), Hyderabad at its power stations at Ramagundam, Farakka, Korba, Vindhyachal, Rihand and Singrauli. These studies have revealed significant environmental gains in the vicinity areas of the project as a result of pursuing sound environment management practices.

Some of these important gains which have been noticed are increase in dense forest area, increase in agriculture area, increase in average rainfall, decrease in waste land etc. In general, the studies, as such, have revealed that there is no significant adverse impact on the ecology due to the project activities in any of these stations. Such studies conducted from time to time around a power project have established comprehensive environment status at various post operational stages of the project.

Geo-hydrological Studies

NTPC has conducted several geohydrological studies of the ash disposal areas at its projects (Singrauli, Rihand, Vindhyachal, Korba, Farakka and Talcher) through reputed institutions like Indian Institutes of Technology, Roorkee; Indian Institutes of Technology, Mumbai, Centre for Studies on Man and Environment, Calcutta.

All these studies conclude that the leaching of heavy metals from ash occurs only under pH 4 or below. In practice, the pH of the ash water is either neutral or alkaline (7 or above) and hence the leaching of heavy metals is highly unlikely.

USE OF WASTE PRODUCTS AND SERVICES -ASH UTILIZATION

Ash is the main solid waste which is put into use for various products and services. NTPC has adopted user friendly policy guidelines on ash utilisation. In order to motivate entrepreneurs to come forward with ash utilisation schemes, NTPC offers several facilities and incentives. These include free issue of all types of ash viz.

Dry Fly Ash/Pond Ash/Bottom Ash and infrastructure facilities, wherever feasible. Necessary help and assistance is also offered to facilitate procurement of land, supply of electricity etc from Government Authorities. Necessary techno-managerial assistance is given wherever considered necessary.

Besides, NTPC uses only ash based bricks and Fly Ash portland pozzolana cement (FAPPC) in most of its construction activities. Demonstration projects are taken up in areas of Agriculture, Building materials, Mine filling etc. The utilisation of ash and ash based products is progressively increasing as a result of the concrete efforts of these groups.

Advanced/Eco-friendly Technologies

NTPC has gained expertise in operation and management of 200 MW and 500 MW Units installed at different Stations all over the country and is looking ahead for higher capacity Unit sizes with super critical steam parameters for higher efficiencies and for associated environmental gains.

At Sipat, higher capacity Units of size of 660 MW and advanced Steam Generators employing super critical steam parameters have already been implemented as a green field project. Higher efficiency Combined Cycle Gas Power Plants are already under operation at all gas-based power projects in NTPC. Advanced clean coal technologies such as Integrated Gasification Combined Cycle (IGCC) have higher efficiencies of the order of 45 per cent as compared to about 38 per cent for conventional plants. NTPC has initiated a techno-economic study under USDOE/USAID for setting up a commercial scale demonstration power plant by using IGCC technology.

These plants can use low grade coals and have higher efficiency as compared to conventional plants. With the massive expansion of power generation, there is also growing awareness among all concerned to keep the pollution under control and preserve the health and quality of the natural environment in the vicinity of the power stations. NTPC is committed to provide affordable and sustainable power in increasingly larger quantity. NTPC is conscious of its role in the national endeavour of mitigating energy poverty, heralding economic prosperity and thereby contributing towards India s emergence as a major global economy.

❐

3

Effects of Global Warming on Ecosystems

Although not every scientist worldwide may look at global warming in the same way, they do overwhelmingly agree that the Earth's atmosphere is getting warmer. Worldwide temperatures have risen more than 1°F over the past century, and 17 of the past 20 years have been the hottest ever recorded. A special report issued by Time magazine on April 3, 2006, the Intergovernmental Panel on Climate Change in their third report, released in 2001, had analysed data from the past two decades representing properties such as air and ocean temperatures and the habitat characteristics and patterns of wildlife. Examples of observed changes included"shrinkage of glaciers, thawing of permafrost, later freezing and earlier breakup of ice on rivers and lakes, lengthening of mid- to high-latitude growing seasons, poleward and altitudinal shifts of plant and animal ranges, declines of some plant and animal populations, and earlier flowering of trees, emergence of insects, and egg- laying in birds. Associations between changes in regional temperatures and observed changes in physical and biological systems have been documented in many aquatic, terrestrial, and marine environments."

The IPCC is an organized group of more than 2,500 climate experts from around the world that consolidates their most recent scientific findings every five to seven years into a single report, which is then presented to the world's political leaders. The IPCC was established in 1988 by the World Meteorological Organization and the United Nations Environment Programme to specifically address the issue of global warming. As a result of their comprehensive analysis, they have determined that this steady warming has had a significant impact on at least 420 animal and plant species and also on natural processes. Furthermore, this has not just occurred in one geographical location but worldwide. In the IPCC's fourth report, released in February 2007, they concluded that it is"very likely" that heat-trapping emissions from human activities have caused "most of the observed increase in globally averaged temperatures since the mid-20th century."

Also, in the February 2007 report, they concluded the following:

- "Human induced warming over recent decades is already affecting many physical and biological processes on every continent. Nearly 90 per cent of the 29,000 observational data series examined revealed changes consistent with the expected response to global warming, and the observed physical and biological responses have been the greatest in the regions that have warmed the most."

In these studies, scientists have been able to break down the natural and human-caused components in order to see how much of an effect humans have had. Human effects can include activities such as burning fossil fuels, agricultural practices, deforestation, industrial processes, the introduction of invasive plant or animal species, and various types of land-use change. In many cases, scientists do not need to look very far to see the effects a warming world is having on the environment and the Earth's ecosystems. Glaciers worldwide are melting at an accelerated rate never seen before. The cap of ice on top of Kilimanjaro is rapidly disappearing, the glaciers of world-renowned Glacier National Park in the United States and Canada are melting and projected to be gone in the next few decades, and the glaciers in the European Alps are experiencing a similar fate. In the world's tropical oceans, vast expanses of beautiful, brilliantly coloured coral reefs are dying off as oceans slowly become too warm.

Unable to survive the higher temperatures, the corals are undergoing a process called bleaching and are turning white and dying. In the Arctic, as temperatures climb, ice is melting at accelerated rates, leaving polar bears stranded, destroying their feeding and breeding grounds, and causing them to starve and drown. Permafrost is melting at accelerated rates. As the ground thaws, it is disrupting the physical and chemical components of the ecosystem by causing the ground to shift and settle, toppling buildings and twisting roads and railroad tracks, as well as releasing methane gas into the atmosphere. Weather patterns are also changing. El Niño events are triggering destructive weather in the eastern Pacific. There has been an increase in extreme weather events, such as hurricanes. Droughts have become more prevalent in some geographical areas, such as parts of Asia, Africa, Australia, and the American Southwest.

Animal and plant habitats have been disrupted, and, as temperatures continue to climb, there have been several documented migrations of individual species moving northward or to higher elevations on individual mountain ranges. Migration patterns are also being affected, such as those already documented of beluga whales, butterflies, and polar bears. Spring is also arriving earlier in some areas, which is now influencing the timing of bird and fish migration, egg laying, leaf unfolding, and spring planting for agriculture. In fact, based on satellite imagery documentation of the Northern Hemisphere, growing seasons have steadily become longer since 1980. While species have been faced with changing environments in the past and have been able to adapt in many cases, the IPCC climate change scientists view this current rate of change with alarm. They fully expect the magnitude of these changes to increase with the temperatures over the next century and beyond.

The concern is that many species and ecosystems will not be able to adapt as rapidly as the effects of global warming will cause the environment to change. In addition, there will also be other disturbances, such as floods, insect infestations, and the spread of disease, wildfire, and drought. Any of these additional challenges can destroy a species or habitat. In particular, alpine and polar species are especially vulnerable to the effects of climate change because as species move northward or higher on mountains, these species' habitats will shrink, leaving them with nowhere to go. With so much evidence, most scientists no longer doubt that global warming is real, nor do they question the fact that humans are to blame.

All it takes is a look at the air quality over significant population and industrial centres to begin to grasp the effect that humans can have on the environment. Based on temperature records kept before the beginning of the Industrial Revolution, carbon dioxide in the atmosphere has increased 30 per cent above those earlier levels. Not only are the levels higher, but they increase annually. The IPCC, at current conditions, by the year 2100, the average temperature is expected to increase between 2° and 11.5°F- an amount more than 50 per cent higher than what was predicted only 50 years ago. Within the IPCC's predicted temperature range, at the lower end, storms would become more frequent and intense, droughts would be more severe, and coastal areas would be flooded by rising sea levels from melting glaciers and ice caps. There would be enough of a disruption that ecosystems worldwide would be thrown out of balance and altered. If, however, the temperature rise falls towards the higher end of the estimate, the results on ecosystems worldwide would be disastrous.

Sea levels could rise so much that entire islands of low elevation, such as the Maldives, could completely disappear. Other areas, such as the Nile Delta and much of the United States' coastal southeast could become completely uninhabitable. Climate zones could shift, completely disrupting land-use practices. For instance, the current agricultural region of the Great Plains in the United States could be shifted to Canada. The southern portion of the United States could become more like Central or South America. Siberia would no longer be a frozen, desolate landscape. Parts of Africa could become dry, desolate wastelands. If this were to happen, it would have a severe impact on the production of agriculture.

Areas currently equipped to produce agriculture would no longer be able to, and areas that were able, based on climate, may not have the financial resources or the proper soils. The ripple effect of these disruptions would be felt worldwide. Millions of people would be forced to migrate from newly uninhabitable regions to new areas where they could survive. This would also affect public health. Rising seas would contaminate freshwater with salt water; there would be more heat-related illnesses and deaths; and disease- carrying rodents and insects, such as mice, rats, mosquitoes, and ticks, would spread diseases such as malaria, encephalitis, Lyme disease, and dengue fever. Scientists of the IPCC agree that one of the most serious aspects of all this drastic change is that it is happening so fast. These changes are happening at a faster pace than the Earth has seen in the last 100 million years. While humans may be able to pick up and move to a new location, animals and their associated ecosystems cannot. The choices people make and the actions they take today will determine the fate of other life and their ecosystems tomorrow.

RESULTS OF GLOBAL WARMING ON ECOSYSTEMS

In a study on the effects of global warming on the Earth's ecosystems conducted by Chris Thomas, a conservation biologist at the University of Leeds in the United Kingdom, he states"Climate change now represents at least as great a threat to the number of species surviving on Earth as habitat destruction and modification." Thomas worked with a group of 18 scientists worldwide in the largest study of its type ever accomplished. The end result of their study came down to this conclusion:"By 2050, rising temperatures, made more severe through human-induced input, such as the burning of fossil fuels, could send more than a million of Earth's land-dwelling plants and animals down the road to extinction." The research team worked by themselves in six biodiversity-rich areas around the world, ranging from Australia to South Africa. As they gathered field data about species distribution and regional climate, they programmed the information into computer climate models.

The purpose of the computer models was to simulate the direction and distance individual species would migrate in response to temperature and climate changes. Once all team members had collected their specific data, they combined it into a single model in order to understand the global concept. Once the model had been carefully evaluated, it was determined that by the year 2050, at predicted global warming rates, 15 to 37 per cent of the 1,103 species studied could be at risk of extinction. When the study area was expanded to cover the entire Earth, the researchers estimated that worldwide more than a million species could begin to face extinction by 2050."This study makes clear that climate change is the biggest new extinction threat," said coauthor Lee Hannah of Conservation International in Washington, D.C."In some cases we found that there will no longer be anywhere climatically suitable for these species to live; in other cases they may be unable to reach distant regions where the climate will be suitable.

Other species are expected to survive in much reduced areas, where they may then be at risk from other threats," said coauthor Guy Midgley of the National Botanical Institute in Cape Town, South Africa."Seeing the range of responses across all 1,103 species, it becomes obvious that we have a lot of work left to do before we can accurately predict what types of animals and plants are most at risk. This range of responses shows that species will not be able to move as whole biological communities, and that the typical natural communities we recognize today will probably not exist under future conditions. Figuring out what will replace them requires a lot of imagination," said coauthor Alison Cameron of the University of Leeds. Chris Thomas, almost all future climate projections expect more warming and even more extinction between 2050 and 2100, and, even though projections are only made to 2100, temperatures will still keep going up and more warming will occur after that.

This group of researchers say that taking action now to slow global warming is important to make sure that climate change ends up on the low end of the prediction in order to avoid"catastrophic extinctions." Thomas also stated that because there may be a time lag between the climate changing and the last individual of a species dying off, the rapid reduction of greenhouse gas emissions may enable some species to survive. It is also important to keep in mind that although some species may be able to migrate successfully to a new location, some plant and animal species that live in high mountain or polar ecosystems cannot move farther to escape warming temperatures. Similarly, coral reef systems cannot just pick up and move to a new location.

Long-established communities have remained where they are in order to survive. Robert Puschendorf, a biologist at the University of Costa Rica, believes these estimates"might be optimistic." As global warming interacts with other factors such as habitat destruction, increase in invasive species, and the buildup of carbon dioxide in the environment, there may be more than a million species that face extinction. In connection with Chris Thomas's study, Richard J. Ladle, et al., from Oxford University"Dangers of Crying Wolf Over Risk of Extinctions," bringing up an issue important to all scientific investigations. It cautioned that the scientific world needs to"learn how to deal with increasingly sensationalist mass media."

They warn that policy-makers must be informed by a"balanced assessment of scientific knowledge and not popular perception created by commercially driven media. Departure from rational objectivity risks undermining public trust in the natural sciences and could play into the hands of antienvironmentalists. This places responsibilities on both scientists and journalists to ensure fair and accurate reporting of their work." Ladle's report clarified that a large portion of the media incorrectly focused on the idea that more than a million species would definitely go extinct by 2050 when the study actually concluded that the extinctions will occur eventually and not in the next 50 years. They reported that 21 out of 29 substances quoting Thomas's study misinformed the public by inferring the species would be extinct by 2050. Their chief concern is that the media, or anyone else, through unclear presentation, could inadvertently increase public cynicism and complacency about climate change and biodiversity loss. The resulting message is that when valuable studies, such as Chris Thomas's, are conducted and the results are released, in order to receive the best results from the scientific community, policy-makers, and the public in general, it is critical not to sensationalize or misrepresent the facts. Otherwise, it puts the real message in jeopardy and may even do more harm than good. Both plant and animal species are at risk due to the effects of global warming.

The World Wildlife Fund, the golden toad and the harlequin frog of Costa Rica have already disappeared as a direct result of global warming. As different components of an ecosystem change, it can upset the natural balance in many ways. For instance, it can disrupt a species by having spring show up a week or two earlier. Over time, delicate balances have been set up between animals and the food they eat. If a certain animal relies on a specific food, but global warming has already caused the food to grow through its life cycle before the animal is ready to use it, then it will have a direct negative effect on that animal-the food will not be available when the animal needs it. That animal's health and existence may be threatened. Furthermore, this could cause a ripple effect in the food chain, having an impact on more than one species.

One example of this is when spring comes earlier than it has in the past. The timing of feeding for newly hatched birds may not correspond to the availability of worms or insects, impacting the fledglings' chances of survival. The WWF, climate records compared with long-term records of flowering and nesting times show a noticeable shift away from each other, depicting global warming trends. In Britain, flowering time and leaf-on records, which date back to 1736, have provided concrete evidence of climate-related seasonal shifts. Long-term trends towards earlier bird breeding, earlier spring migrant arrival, and later autumn departure dates have also been recorded in North America. Changes in migratory patterns have also been documented in Europe.

Once again, the WWF documents that climate change and global warming are currently affecting species in many ways. Animals and plants that require cooler temperatures in order to survive, which need to either migrate northward or higher in elevation in a mountain ecosystem, are already being documented in several places worldwide. This is occurring in the European Alps, in Queensland in Australia, and in the rain forests of Costa Rica. Fish in the North Sea have been documented migrating northward. Fish populations that used to inhabit areas around Cornwall, England, have migrated as far north as the Shetland and Orkney Islands. WWF global warming experts believe, based on this evidence, that"the impacts on species are becoming so significant that their movements can be used as an indicator of a warming world.

They are silent witnesses of the rapid changes being inflicted on the Earth. In fact, these same scientists believe that global warming could begin causing extinctions of animal species in the near future because the heating caused by accelerated global warming has a severe impact on the Earth's many delicate ecosystems-both on the land and the species that live on it. Worldwide, there are species and habitats that have now been identified as being threatened and endangered due to the effects of global warming. Because ecosystems can be altered to the point where the damage becomes irreversible and species must either adapt to survive or face extinction, it is critical that the issue of global warming be addressed and acted upon now before it is too late.

OBSERVED AND EXPECTED EFFECTS ON ECOSYSTEMS

There are several pieces of physical evidence that scientists have already identified indicating that global warming is already in progress and affecting all the ecosystems on Earth. By monitoring the health of ecological conditions, scientists can see the effects climate change is having on the individual components that comprise the ecosystem. Because an ecosystem is such a tightly knit system of living things within their natural environment, if one component is affected, a ripple effect can be started, eventually endangering the entire ecosystem.

Polar And Ice-Related Changes

Of all the Earth's ecosystems, climate change in the polar regions is expected to be more rapid and more severe than anywhere else. If snow and ice are melted, this will greatly change the albedo of the environment.

As darker surfaces increase, more sunlight will be absorbed, rapidly heating up the Earth's surface and atmosphere. In addition, as worldwide species continue to migrate northward under warming temperatures, thick dark vegetation will crowd areas that were once wideopen snowfields, also lowering the albedo. The IPCC, the average annual temperature in the Arctic has increased by 1.67°F over the past century—which equals a rate roughly twice as fast as the global average. Winter temperatures have been consistently 3.3°F warmer over the past century. The effects of this warming have been seen in decreases in thickness and extent of sea ice, the melting of permafrost, and later freezing and earlier breakup dates of winter sea ice.

Glaciers worldwide are also melting: There are glaciers on all the Earth's continents except Australia and at all latitudes from the Tropics to the polar regions. There is widespread evidence that glaciers are retreating in many areas of the world. Because sea ice in the polar regions is breaking up earlier in the year, polar bears and walrus are already suffering. Their feeding and breeding grounds are disappearing, their territorial boundaries are gone, and they are dying. Early breakup is also affecting the hunting habits and lifestyles of northern native inhabitants, forcing many to abandon lifelong rich, traditional cultures and relocate to other areas.

Fire and Drought

Over large areas of the Earth, nights have warmed up more than days have. In fact, since 1950, minimum temperatures on land have rapidly increased. An increase in warm temperatures will lead to increases in the number of heat waves that strike urban areas, which will cause more heat-related illnesses and deaths. Global warming is causing a more intense hydrologic cycle with increased evaporation. The greater the evaporation rates, the more soils and vegetation will dry out. As temperatures rise and vegetation dries out, areas will become drier under droughtlike conditions, and wildfires will become more common. This occurred during the 2007 summer and fall in California. During this tragic event, more than 772 square miles of land burned from Santa Barbara County to the U.S.-Mexico border. Nearly 100 people were injured, 9 died, and more than 1,500 homes were destroyed. Wildfires forced 265,000 residents to flee their homes.

Biological Changes

Global warming also affects the occurrence and spread of disease. It makes large populations vulnerable if the pathogen is spread quickly. Warmer temperatures and more precipitation will help spread disease organisms from rodents and insects to larger areas. The world's undeveloped countries are expected to be hit the hardest. Rising sea levels due to the melting of glaciers and ice caps are causing coral to lose the symbiotic algae that they must have for nutrition. This algae is also what gives them their beautiful, vibrant colours. When the algae die, the coral looks white and is referred to as bleached.

It only takes a small increase in temperature above normal summer levels for periods of time as short as only two or three days to cause this reaction. In 1998, one of the hottest years on record, coral reefs worldwide experienced the most extensive bleaching ever recorded. Coral bleaching was reported in 60 countries and island nations in the Indian Ocean, Red Sea, Pacific Ocean, Persian Gulf, Mediterranean, and Caribbean. Plant and animal species are located where the climatic factors enable them to thrive. If any of these characteristics change-as they do with global warming-then species will attempt to migrate. Whether they are successful or not depends on several factors: the rate of change, availability of acceptable habitat, a physical way to relocate to acceptable habitat, and avoidance of predators.

If any of these factors works out wrong, the species can become threatened, endangered, or extinct. NASA, based on data obtained through satellite observation over the past 20 years, areas in both North America and Eurasia have longer growing seasons associated with the buildup of greenhouse gases in the atmosphere. During the course of data collection, NASA scientists noticed dramatic changes in the timing of when leaves first appeared in the spring and when they fell off in the fall. By monitoring when things turn green, NASA scientists determined that in Eurasia the growing season is currently almost 18 days longer than it was two decades ago. Today, spring arrives a week earlier and autumn 10 days later than they did in the past.

Physical Changes

The rise in mean global surface temperature has caused spring to come earlier in many parts of the world. This has led to a longer growing season in middle and high-latitude areas. The effects of this are widespread-leaves come out earlier and stay longer, and breeding and migration patterns of wildlife are affected. For instance, in the northeastern United States, the frost-free season now begins approximately 11 days earlier than it did in the 1950s. As worldwide temperatures climb, the hydrologic cycle will intensify, producing more intense phenomenon such as flooding, landslides, and erosion. The areas at highest risk are those at mid to high latitudes. Trends are already apparent in several polar locations in the Northern Hemisphere. P.Y. Groisman and D. R. Easterling of the NCDC, over the past few decades snowfall has increased about 20 per cent over northern Canada and about 11 per cent over Alaska. An increase in snowfall has also been observed over China. T. R. Karl and R. W. Knight, observations for the last century indicate that extreme weather events in the United States have increased by about 20 per cent. Increases in heavy rainfall have also been reported in Japan and northeastern Australia.

Tidal gauges are placed worldwide and mean sea level is monitored. Over the past century, global mean sea level has risen 4 to 10 inches with the average being 7 inches. This rate is greater than what the average has been over the past few thousand years. The IPCC, although unsure how much ocean levels will rise in the next century, has projected that the rate will be at least two to four times the rate of the last century. The reason for the uncertainty is that the behaviour of the Antarctic and Greenland ice sheets remains uncertain at this point. If their melting rate increases, future sea-level rise will most likely be on the larger side of the projection. Sea-level rise and coastal flooding are also governed by wind and pressure patterns, ocean circulation, and the characteristics of the coastline-whether there are coastal wetlands, beaches, islands, or other structures to act as barriers.

ECOSYSTEM SERVICES IN GLOBAL WARMING

Humankind benefits from a multitude of resources and processes that are supplied by natural ecosystems. Collectively, these benefits are known as ecosystem services and include products like clean drinking water and processes such as the decomposition of wastes. While scientists and environmentalists have discussed ecosystem services for decades, these services were popularized and their definitions formalized by the United Nations 2004 Millennium Ecosystem Assessment, a four-year study involving four broad categories: provisioning, such as the production of food and water; regulating, such as the control of climate and disease; supporting, such as nutrient cycles and crop pollination; and cultural, such as spiritual and recreational benefits.

As human populations grow, so do the resource demands imposed on ecosystems and the impacts of our global footprint. Natural resources are not invulnerable and infinitely available. The environmental impacts of anthropogenic actions, which are processes or materials derived from human activities, are becoming more apparent – air and water quality are increasingly compromised, oceans are being overfished, pests and diseases are extending beyond their historical boundaries, and deforestation is exacerbating flooding downstream.

It has been reported that approximately 40-50 per cent of Earth's ice-free land surface has been heavily transformed or degraded by anthropogenic activities, 66 per cent of marine fisheries are either overexploited or at their limit, atmospheric CO_2 has increased more than 30 per cent since the advent of industrialization, and nearly 25 per cent of Earth's bird species have gone extinct in the last two thousand years. Society is increasingly becoming aware that ecosystem services are not only limited, but also that they are threatened by human activities. The need to better consider long-term ecosystem health and its role in enabling human habitation and economic activity is urgent. To help inform decision-makers, many ecosystem services are being assigned economic values, often based on the cost of replacement with anthropogenic alternatives. The ongoing challenge of prescribing economic value to nature, for example through biodiversity banking, is prompting transdisciplinary shifts in how we recognize and manage the environment, social responsibility, business opportunities, and our future as a species.

The simple notion of human dependence on Earth's ecosystems probably reaches to the start of our species' existence, when we benefited from the products of nature to nourish our bodies and for shelter from harsh climates. Recognition of how ecosystems could provide more complex services to mankind date back to at least Plato who understood that deforestation could lead to soil erosion and the drying of springs.

However, modern ideas of ecosystem services probably began with Marsh in 1864 when he challenged the idea that Earth's natural resources are not infinite by pointing out changes in soil fertility in the Mediterranean. Unfortunately, his observations and cautions passed largely unnoticed at the time and it was not until the late 1940s that society's attention was again brought to the matter. During this era, three key authors – Osborn, Vogt, and Leopold – awakened and promoted recognition of human dependence on the environment with the idea of 'natural capital'.

In 1956, Sears drew attention to the critical role of the ecosystem in processing wastes and recycling nutrients. An environmental science textbook called attention to "the most subtle and dangerous threat to man's existence... the potential destruction, by man's own activities, of those ecological systems upon which the very existence of the human species depends".

The term 'environmental services' was finally introduced in a report of the Study of Critical Environmental Problems, which listed services including insect pollination, fisheries, climate regulation and flood control. In following years, variations of the term were used, but eventually 'ecosystem services' became the standard in scientific literature.

Modern expansions of the ecosystem services concept include socio-economic and conservation objectives, which are discussed below. For a more complete history of the concepts and terminology of ecosystem services.

Ecology

Understanding of ecosystem services requires a strong foundation in ecology, which describes the underlying principles and interactions of organisms and the environment. Since the scales at which these entities interact can vary from microbes to landscapes, milliseconds to millions of years, one of the greatest remaining challenges is the descriptive characterization of energy and material flow between them.

For example, the area of a forest floor, the detritus upon it, the microorganisms in the soil and characteristics of the soil itself will all contribute to the abilities of that forest for providing ecosystem services like carbon sequestration, water purification, and erosion prevention to other areas within the watershed. Note that it is often possible for multiple services to be bundled together and when benefits of targeted objectives are secured, there may also be ancillary benefits – the same forest may provide habitat for other organisms as well as human recreation, which are also ecosystem services.

The complexity of Earth's ecosystems poses a challenge for scientists as they try to understand how relationships are interwoven among organisms, processes and their surroundings. As it relates to human ecology, a suggested research agenda for the study of ecosystem services includes the following steps:

1. Identification of ecosystem service providers – species or populations that provide specific ecosystem services – and characterization of their functional roles and relationships;
2. Determination of community structure aspects that influence how ESPs function in their natural landscape, such as compensatory responses that stabilize function and non-random extinction sequences which can erode it; more than 1300 scientists worldwide. This grouped ecosystem services into of services;
3. Measurement of the spatial and temporal scales ESPs and their services operate on.

Recently, a technique has been developed to improve and standardize the evaluation of ESP functionality by quantifying the relative importance of different species in terms of their efficiency and abundance. Such parametres provide indications of how species respond to changes in the environment and are useful for identifying species that are disproportionately important at providing ecosystem services.

However, a critical drawback is that the technique does not account for the effects of interactions, which are often both complex and fundamental in maintaining an ecosystem and can involve species that are not readily detected as a priority.

Even so, estimating the functional structure of an ecosystem and combining it with information about individual species traits can help us understand the resilience of an ecosystem amidst environmental change.

Many ecologists also believe that the provision of ecosystem services can be stabilized with biodiversity. Increasing biodiversity also benefits the variety of ecosystem services available to society. Understanding the relationship between biodiversity and an ecosystem's stability is essential to the management of natural resources and their services.

The Redundancy Hypothesis

The concept of ecological redundancy is sometimes referred to as functional compensation and assumes that more than one species performs a given role within an ecosystem. More specifically, it is characterized by a particular species increasing its efficiency at providing a service when conditions are stressed in order to maintain aggregate stability in the ecosystem.

However, such increased dependence on a compensating species places additional stress on the ecosystem and often enhances its susceptibility to subsequent disturbance. The redundancy hypothesis can be summarized as "species redundancy enhances ecosystem resilience".

The Rivet Hypothesis

Another idea uses the analogy of rivets in an airplane wing to compare the exponential effect the loss of each species will have on the function of an ecosystem; this is sometimes referred to as rivet popping. If only one species disappears, the loss of the ecosystem's efficiency as a whole is relatively small; however if several species are lost, the system essentially collapses as an airplane wing would, were it to lose too many rivets.

The hypothesis assumes that species are relatively specialized in their roles and that their ability to compensate for one another is less than in the redundancy hypothesis. As a result, the loss of any species is critical to the performance of the ecosystem. The key difference is the rate at which the loss of species affects total ecosystem function.

The Portfolio Effect

A third explanation, known as the portfolio effect, compares biodiversity to stock holdings, where diversification minimizes the volatility of the investment, or in this case, the risk in stability of ecosystem services. This is related to the idea of response diversity where a suite of species will exhibit differential responses to a given environmental perturbation and therefore when considered together, they create a stabilizing function that preserves the integrity of a service.

Several experiments have tested these hypotheses in both the field and the lab. In ECOTRON, a laboratory in the UK where many of the biotic and abiotic factors of nature can be simulated, studies have focused on the effects of earthworms and symbiotic bacteria on plant roots. These laboratory experiments seem to favour the rivet hypothesis. However, a study on grasslands at Cedar Creek Reserve in Minnesota seems to support the redundancy hypothesis, as have many other field studies.

Economics

There is extensive disagreement regarding the environmental and economic values of ecosystem services. Some people may be unaware of the environment in general and humanity's interrelatedness with the natural environment, which may cause misconceptions. Although environmental awareness is rapidly improving in our contemporary world, ecosystem capital and its flow are still poorly understood, threats continue to impose, and we suffer from the so-called 'tragedy of the commons'.

Many efforts to inform decision-makers of current versus future costs and benefits now involve organizing and translating scientific knowledge to economics, which articulate the consequences of our choices in comparable units of impact on human well-being. An especially challenging aspect of this process is that interpreting ecological information collected from one spatial- temporal scale does not necessarily mean it can be applied at another; understanding the dynamics of ecological processes relative to ecosystem services is essential in aiding economic decisions.

Weighting factors such as a service's irreplaceability or bundled services can also allocate economic value such that goal attainment becomes more efficient.The economic valuation of ecosystem services also involves social communication and information, areas that remain particularly challenging and are the focus of many researchers. In general, the idea is that although individuals make decisions for any variety of reasons, trends reveal the aggregative preferences of a society, from which the economic value of services can be inferred and assigned.

The six major methods for valuing ecosystem services in monetary terms are:

1. Avoided cost: Services allow society to avoid costs that would have been incurred in the absence of those services (e.g. waste treatment by wetland habitats avoids health costs)
2. Replacement cost: Services could be replaced with man-made systems (e.g. restoration of the Catskill Watershed cost less than the construction of a water purification plant)
3. Factor income: Services provide for the enhancement of incomes (e.g. improved water quality increases the commercial take of a fishery and improves the income of fishers)
4. Travel cost: Service demand may require travel, whose costs can reflect the implied value of the service (e.g. value of ecotourism experience is at least what a visitor is willing to pay to get there)
5. Hedonic pricing: Service demand may be reflected in the prices people will pay for associated goods (e.g. coastal housing prices exceed that of inland homes)
6. Contingent valuation: Service demand may be elicited by posing

hypothetical scenarios that involve some valuation of alternatives (e.g. visitors willing to pay for increased access to national parks)

Management and Policy

Utility of Radiative Forcing: The TAR and other assessments have concluded that RF is a useful tool for estimating, to a first order, the relative global climate impacts of differing climate change mechanisms. In particular, RF can be used to estimate the relative equilibrium globally averaged surface temperature change due to different forcing agents. However, RF is not a measure of other aspects of climate change or the role of emissions. Previous GCM studies have indicated that the climate sensitivity parameter was more or less constant between mechanisms.

However, this level of agreement was found not to hold for certain mechanisms such as ozone changes at some altitudes and changes in absorbing aerosol. Because the climate responses, and in particular the equilibrium climate sensitivities, exhibited by GCMs vary by much more than 25 per cent, Ramaswamy and Jacob concluded that RF is the most simple and straightforward measure for the quantitative assessment of climate change mechanisms, especially for the LLGHGs. This part discusses the several studies since the TAR that have examined the relationship between RF and climate response. Note that this assessment is entirely based on climate model simulations.

Vertical Forcing Patterns and Surface Energy Balance Changes

The vertical structure of a forcing agent is important both for efficacy and for other aspects of climate response, particularly for evaluating regional and vertical patterns of temperature change and also changes in the hydrological cycle. For example, for absorbing aerosol, the surface forcings are arguably a more useful measure of the climate response than the RF. It should be noted that a perturbation to the surface energy budget involves sensible and latent heat fluxes besides solar and longwave irradiance; therefore, it can quantitatively be very different from the RF, which is calculated at the tropopause, and thus is not representative of the energy balance perturbation to the surface-troposphere system.

While the surface forcing adds to the overall description of the total perturbation brought about by an agent, the RF and surface forcing should not be directly compared nor should the surface forcing be considered in isolation for evaluating the climate response. Therefore, surface forcings are presented as an important and useful diagnostic tool that aids understanding of the climate response.

Alternative Methods of Calculating Radiative Forcing

RFs are increasingly being diagnosed from GCM integrations where the calculations are complex. This stage also discusses several mechanisms that include some response in the troposphere, such as cloud changes. These mechanisms are not initially radiative in nature, but will eventually lead to a radiative perturbation of the surface-troposphere system that could conceivably be measured at the TOA. Jacob refer to these mechanisms as non-radiative forcings. Alternatives to the standard stratospherically adjusted RF definition have been proposed that may help account for these processes.

Since the TAR, several studies have employed GCMs to diagnose the zero- surface-temperature-change RF. These studies have used a number of different methodologies. Shine fixed both land and sea surface temperatures globally and calculated a radiative energy imbalance: this technique is only feasible in GCMs with relatively simple land surface parametrizations. Hansen fixed sea surface temperatures and calculated an RF by adding an extra term to the radiative imbalance that took into account how much the land surface temperatures had responded.

Sokolov diagnosed the zero-surface-temperaturechange RF by computing surface-only and atmospheric-only components of climate feedback separately in a slab model and then modifying the stratospherically adjusted RF by the atmospheric-only feedback component. Gregory used a regression method with a globally averaged temperature change ordinate to diagnose the zero- surface-temperature-change RF: this method had the largest uncertainties. Shine, Hansen and Sokolov all found that that the fixed-surface-temperature RF was a better predictor of the equilibrium global mean surface temperature response than the stratospherically adjusted RF. Further, it was a particularly useful diagnostic for changes in absorbing aerosol where the stratospherically adjusted RF could fail as a predictor of the surface temperature response.

Differences between the zero-surface-temperature-change RF and the stratospherically adjusted RF can be caused by semi-direct and cloud-aerosol interaction effects beyond the cloud albedo RF. For most mechanisms, aside from the case of certain aerosol changes, the difference is likely to be small.

These calculations also remove problems associated with defining the tropopause in the stratospherically adjusted RF definition. However, stratospherically adjusted RF has the advantage that it does not depend on relatively uncertain components of a GCM's response, such as cloud changes. For the LLGHGs, the stratospherically adjusted RF also has the advantage that it is readily calculated in detailed off-line radiation codes. However, to first order, all methods are comparable and all prove useful for understanding climate response.

Linearity of the Forcing—Response Relationship

Reporting findings from several studies, the TAR concluded that responses to individual RFs could be linearly added to gauge the global mean response, but not necessarily the regional response. Since then, studies with several equilibrium and/or transient integrations of several different GCMs have found no evidence of any non-linearity for changes in greenhouse gases and sulphate aerosol. Two of these studies also examined realistic changes in many other forcing agents without finding evidence of a non-linear response. In all four studies, even the regional changes typically added linearly.

However, Meehl observed that neither precipitation changes nor all regional temperature changes were linearly additive. This linear relationship also breaks down for global mean temperatures when aerosol-cloud interactions beyond the cloud albedo RF are included in GCMs. Studies that include these effects modify clouds in their models, producing an additional radiative imbalance.

Rotstayn and Penner found that if these aerosol-cloud effects are accounted for as additional forcing terms, the inference of linearity can be restored. Studies also find non-linearities for large negative RFs, where static stability changes in the upper troposphere affect the climate feedback.

For the magnitude and range of realistic RFs discussed in this stage, and excluding cloud-aerosol interaction effects, there is high confidence in a linear relationship between global mean RF and global mean surface temperature response.

Efficacy and Effective Radiative Forcing

Efficacy is defined as the ratio of the climate sensitivity parameter for a given forcing agent to the climate sensitivity parameter for CO_2 changes, that is, Ei = $\lambda i / \lambda CO_2$. Efficacy can then be used to define an effective RF. For the effective RF, the climate sensitivity parameter is independent of the mechanism, so comparing this forcing is equivalent to comparing the equilibrium global mean surface temperature change. That is, $\Delta Ts = \lambda CO_2 \times Ei \times RFi$ Preliminary studies have found that efficacy values for a number of forcing agents show less model dependency than the climate sensitivity values.

Effective RFs have been used get one step closer to an estimator of the likely surface temperature response than can be achieved by using RF alone. Adopting the zero-surface-temperature-change RF, which has efficacies closer to unity, may be another way of achieving similar goals.

This part assesses the efficacy associated with stratospherically adjusted RF, as this is the definition of RF adopted in this stage. Therefore, cloudaerosol interaction effects beyond the cloud albedo RF are included in the efficacy term. As space is limited not all these studies are explicitly discussed in the main text.

Generic Understanding

Since the TAR, several GCM studies have calculated efficacies and a general understanding is beginning to emerge as to how and why efficacies vary between mechanisms. The initial climate state, and the sign and magnitude of the RF have less importance but can still affect efficacy.

These studies have also developed useful conceptual models to help explain variations in efficacy with forcing mechanism. The efficacy primarily depends on the spatial structure of the forcings and the way they project onto the various different feedback mechanisms. Therefore, different patterns of RF and any non-linearities in the forcing response relationship affects the efficacy. Nearly all studies that examine it find that high-latitude forcings have higher efficacies than tropical forcings. Efficacy has also been shown to vary with the vertical distribution of an applied forcing.

Forcings that predominately affect the upper troposphere are often found to have smaller efficacies compared to those that affect the surface. However, this is not ubiquitous as climate feedbacks will depend on the static stability of the troposphere and hence the sign of the temperature change in the upper troposphere.

Long-Lived Greenhouse Gases

The few models that have examined efficacy for combined LLGHG changes generally find efficacies slightly higher than 1.0. Further, the most recent result from the NCAR Community Climate Model GCM indicates an efficacy of over 1.2 with no clear reason of why this changed from earlier versions of the same model. Individual LLGHG efficacies have only been analysed in two or three models. Two GCMs suggest higher efficacies from individual components. In contrast another GCM gives efficacies for CFCs and CH_4 that are slightly less than one. Overall there is medium confidence that the observed changes in the combined LLGHG changes have an efficacy close to 1.0 but there are not enough studies to constrain the efficacies for individual species.

Solar

Solar changes, compared to CO_2, have less high-latitude RF and more of the RF realised at the surface. Established but incomplete knowledge suggests that there is partial compensation between these effects, at least in some models, which leads to solar efficacies close to 1.0. All models with a positive solar RF find efficacies of 1.0 or smaller. One study finds a smaller efficacy than other models. However, their unique methodology for calculating climate sensitivity has large uncertainties.

These studies have only examined solar RF from total solar irradiance change; any indirect solar effects are not included in this efficacy estimate.

Overall, there is medium confidence that the direct solar efficacy is within the 0.7 to 1.0 range.

Ozone

Stratospheric ozone efficacies have normally been calculated from idealised ozone increases. Experiments with three models found higher efficacies for such changes; these were due to larger than otherwise tropical tropopause temperature changes which led to a positive stratospheric water vapour feedback. However, this mechanism may not operate in the two versions of the GISS model, which found smaller efficacies. Only one study has used realistic stratospheric ozone changes; thus, knowledge is still incomplete. Conclusions are only drawn from the idealised studies where there is medium confidence that the efficacy is within a 0.5 to 2.0 range and established but incomplete physical understanding of how and why the efficacy could be larger than 1.0. There is medium confidence that for realistic tropospheric ozone perturbations the efficacy is within the 0.6 to 1.1 range.

Scattering Aerosol

For idealised global perturbations, the efficacy for the direct effect of scattering aerosol is very similar to that for changes in the solar constant. As for ozone, realistic perturbations of scattering aerosol exhibit larger changes at higher latitudes and thus have a higher efficacy than solar changes. Although the number of modelling results is limited, it is expected that efficacies would be similar to other solar effects; thus there is medium confidence that efficacies for scattering aerosol would be in the 0.7 to 1.1 range. Efficacies are likely to be similar for scattering aerosol in the troposphere and stratosphere. The efficacy of the cloud albedo RF accounts for cloud lifetime effects. Only two studies contained enough information to calculate efficacy in this way and both found efficacies higher than 1.0. However, the uncertainties in quantifying the cloud lifetime effect make this efficacy very uncertain. If cloud lifetime effects were excluded from the efficacy term, the cloud albedo efficacy would very likely be similar to that of the direct effect.

Absorbing Aerosol

For absorbing aerosols, the simple ideas of a linear forcingresponse relationship and efficacy can break down. Aerosols within a particular range of single scattering albedos have negative RFs but induce a global mean warming, that is, the efficacy can be negative. The surface albedo and height of the aerosol layer relative to the cloud also affects this relationship.

Studies that increase BC in the planetary boundary layer find efficacies much larger than 1.0. These studies also find that efficacies are considerably smaller than 1.0 when BC aerosol is changed the boundary layer. These changes in efficacy are at least partly attributable to a semi-direct effect whereby absorbing aerosol modifies the background temperature profile and tropospheric cloud. Another possible feedback mechanism is the modification of snow albedo by BC aerosol; however, this report does not classify this as part of the response, but rather as a separate RF.

Most GCMs likely have some representation of the semi-direct effect but its magnitude is very uncertain and dependent on aspects of cloud parametrizations within GCMs. Two studies using realistic vertical and horizontal distributions of BC find that overall the efficacy is around 0.7. However, Hansen acknowledge that they may have underestimated BC within the boundary layer and another study with realistic vertical distribution of BC changes finds an efficacy of 1.3. Further, Penner also modelled BC changes and found efficacies very much larger and very much smaller than 1.0 for biomass and fossil fuel carbon, respectively found similar efficacies for biomass and fossil fuel carbon). In summary there is no consensus as to BC efficacy and this may represent problems with the stratospherically adjusted definition of RF.

Other Forcing Agents

Efficacies for some other effects have been evaluated by one or two modelling groups. Hansen found that land use albedo RF had an efficacy of roughly 1.0, while the BCsnow albedo RF had an efficacy of 1.7. Ponater found an efficacy of 0.6 for contrail RF and this agrees with a suggestion from Hansen that high-cloud changes should have smaller efficacies. The results of Hansen and Forster and Shine suggest that stratospheric water vapour efficacies are roughly one.

Efficacy And The Forcing-Response Relationship

Efficacy is a new concept introduced since the TAR and its physical understanding is becoming established. When employing the stratospherically adjusted RF, there is medium confidence that efficacies are within the 0.75 to 1.25 range for most realistic RF mechanisms aside from aerosol and stratospheric ozone changes. There is medium confidence that realistic aerosol and ozone changes have efficacies within the 0.5 to 2.0 range. Further, zero- surface-temperature-change RFs are very likely to have efficacies significantly closer to 1.0 for all mechanisms.

BASIC COMPONENTS OF THE ECOSYSTEM

It is helpful to visualize ecosystems as consisting of four basic components: abiotic substances, producer organisms, consumer organisms and decomposer organisms.

Abiotic Components

Abiotic substances are the inorganic and organic substances not momentarily present in living organisms. These include water, carbon dioxide, oxygen, nitrogen, minerals, salts, acids, bases and the entire range of elements and compounds outside living organisms at any given point in time. Many elements may be tightly bound in inorganic compounds, such as silicon in sandstone or aluminum in feldspar, and are unavailable to living organisms. Elements which are readily available to living organisms such as free O_2 or CO_2, or they may be in an inaccessible form such as silicon dioxide (SiO_2) in quartz, a major component of granite. Similarly, potassium may be readily available to plants in the form of KCL in soil, but relatively unavailable in the form of $KAlSi_3O_8$ in orthoclase or monoclinic feldspar, one of the commonest of all minerals.

An important property of an ecosystem which determines its productivity is the form and composition in which bioactive elements and compounds occur. For example, an ecosystem may have a substantial abundance of vital nutrients, such as nitrates and phosphates, but if they are present in relatively insoluble particulate form as they would be if linked to ferric ions, they would not be so readily available to plants as if they were in the soluble form of potassium or calcium nitrate and phosphate. One of the most important qualities of an ecosystem is the rate of release of nutrients from solids, for this regulates the rate of function of the entire system.

Producers

Producer organisms are bacteria and plants which synthesize organic compounds. They are said to be autotrophic or self-productive, in that they take inorganic compounds and manufacture organic materials and living protoplasm from them. All green plants, including microscopic algae, are producer organisms since they exhibit photosynthesis, and some bacteria are producers since they may exhibit chemosynthesis or photosynthesis. Obviously, all life depends upon the basic productive capacity of green plants and bacteria.

Consumers

Consumer organisms are animals which utilize the organic materials directly or indirectly manufactured by plants. Consumers are unable to produce their own organic compounds for basic nutritive purposes. They are said to be heterotrophic, which means different or varied in nutritional source. Primary consumers or herbivores directly consume the organic compounds of plants. Secondary consumers may be omnivores or carnivores which depend partially or entirely on other animals for food. Tertiary and quartenary consumers may be the second or third-stage predator, for example, a hawk feeding on a weasel which in turn consumed a mouse.

Decomposers

Decomposer organisms are bacteria and fungi which degrade organic compounds. Their nutrition is said to be saprophytic, that is, associated with rotten an ecosystem—they reduce the complex organic molecules of dead plants and animals to simpler organic compounds which can be absorbed by green plants as vital nutrients. They provide the final essential link in the cycle of life. They are necessary for the renewal of life, for if decomposers were not active, organic compounds would become locked into complex insoluble molecules which could not be utilized as nutrients by plants.

Ecosystems involve, of course, a wide variety of life forms not specifically mentioned in the preceding paragraphs, but virtually all components of an ecosystem can be classified into producers, consumers, or decomposers. For example, parasites are merely specialized consumers. Plant parasites feed directly on plants and are thus herbivores; animal parasites derive their nutrition from other animals, and are thus carnivores differing from predators only in the fact that they normally do not kill the host. Scavengers such as vultures are also carnivores, differing from predators by the fact that they feed on an animal after it has died from some other cause.

Incomplete Ecosystems

Almost all ecosystems have all four basic components discussed above, though in some cases it is possible for incomplete ecosystems to exist. These are ecosystems lacking one or more basic components.

An example of an incomplete ecosystem lacking producers is the abyssal depths of the sea where only consumers and decomposers exist. In the realm of complete darkness green plants cannot survive. Scavengers and decomposers live on the fall-out of animals, plants and organ matter from the upper layers of the ocean. Predators might also be present to feed upon the scavengers. Hence, the ecosystem depends on extrinsic production, namely, the fall-out from upper levels. It might be possible, of course, for a few chemosynthetic bacteria to be present, but they would not produce a significant volume of organic material.

The same situation exists in caves where complete darkness prevents the growth of green plants. Again, a few chemosynthetic bacteria might be present, but they would not produce a significant amount of organic material. Practically all cave-dwelling-animals must depart from the cave, as do bats, or depend on extrinsically produced nutrients which enter the cave by flowing water or seepage.

The central core of the city might also be considered an incomplete ecosystem without producers, at least from the human standpoint. Some green plants obviously exist, but they would not supply meaning ful production for humans, sparrows, dogs, cats, etc. For all of these, the inner city requires extrinsic production and imported food. The only other alternative is for the inhabitants to leave the inner city and feed in peripheral areas. This undoubtedly occurs with starlings and pigeons, creating an ecological situation analogous to bats in a cave. The lack of production in an inner city is not due to a lack of light, but to a lack of soil and suitable substrate.

In other ways, cities may be considered incomplete ecosystems, ecologically parasitic upon the surrounding landscape. Not only do they import food, but they must also import fresh air and water. At the same time, they must export waste products—sewage, solid waste, carbon dioxide, sulphar dioxide, etc. If cities were encapsulated from their surrounding environments they would soon perish from thirst, starvation, asphyxiation, or the accumulation of waste products. In exchange for this life support, cities, of course, provide a great many economic and cultural benefits—jobs, housing, transportation, manufacturing, education, etc. So the relationship between city and landscape is vital in both directions, but it is particularly important to remember, as cities expand, that they cannot sustain themselves.

Incomplete ecosystems also exist in specialized cases where producers and decomposers are present without consumers. A theoretical example would be a massive bloom of some toxic algae in an aquatic ecosystem, where the algae would create toxic conditions for zooplankton and fish and all other possible consumers. Then a process of excessive production and massive decomposition would go hand in hand. This would be a highly unstable and undesirable circumstance, but it has been known to occur, as, for example, in the red tides of Florida. A third type of incomplete ecosystem might even be called an abiotic ecosystem, that is, one without living organisms, a self-contradiction in terms. They should more properly be called abiotic environments. Apollo space flights have shown so far that the moon is abiotic. Local areas on earth may be abiotic, for example, the high altitude ice plateau of Antarctica is probably devoid of living organisms over rather extensive areas. Closer home for most of us, the Copperhill basin of Tennessee is an area devoid of life, where fumes from copper smelters are toxic to all organisms within a certain downwind area. Possibly a few bacteria exist in very limited places, but for all practical purposes no plants or animals can survive.

GLOBAL WARMING STRESS TO CORAL REEFS

The World Wildlife Fund (WWF) reports that coral reefs around the world have been severely damaged by unusually warm ocean waters. They predict that less than 5 per cent of Australia's Great Barrier Reef will remain by 2050 if the world fails to reduce CO_2 emissions. They project that "if the present rate of destruction continues, a good proportion of the world's coral reefs could be killed within our lifetime." WWF has identified both the Seychelles Islands and American Samoa as locations under high stress for coral bleaching. The United Nations Educational, Scientific and Cultural Organization (UNESCO) recognizes the Seychelles Islands as a natural World Heritage Site.

They have a high diversity of coral and support rare land species, such as the giant tortoise. In addition to increasing ocean temperatures, these areas are also threatened by global warming because of more frequent tropical storms (which could break up the coral) and more frequent rains, flooding, and river run-off (which deposits sediments in the ocean). The National Aeronautics and Space Administration (NASA), reef habitats are so complex and warrant so much exploration, "that for marine biologists, the destruction of the reefs has proven to be as frustrating as it is heart breaking. At the rate the reefs are disappearing, they may be beyond repair by the time a comprehensive plan to save reefs can be put into place." Scientists at NASA's Goddard Space Flight Centre and at several universities around the world have at least a partial solution. They have been interpreting satellite imagery of the world's oceans obtained from Landsat 7 and other high-resolution sensors to map the shallow waters around the ocean's margins, creating maps of the reef environments.

With funding, they expect to soon have a comprehensive map of the world's reef systems that they can use to identify large-scale threats to the reefs. In the meantime, Abdul Azeez Abdul Hakeem, a Maldivian scientist, is dedicated to helping corals on Maldives—a nation of 1,200 islands in the Indian Ocean—survive global warming. Azeez is the director of conservation for an eco-friendly resort called Banyan Tree Maldives. Besides being a premier vacation resort, it also hosts a worldclass marine laboratory managed by Azeez. The staff spends most of their time studying the coral and maintaining the ecosystem—the reefs provide homes for many exotic fish and attract travellers from all over the world. Azeez became involved with coral in 1998 when a strong El Niño warmed the ocean and put the coral at risk. "It rose to about 91°F (33°C), so 33 was boiling hell, and about 80 to 90 per cent of the corals in the Maldives died. I never believed that an entire region could be wiped out. No one believed that this could happen until it hit us. Then only I also realised yes we are in danger because of global warming and this can happen again and again." When that happened, Azeez began looking for ways to protect the coral.

He also knew that corals on an artificial reef nearby had survived the 1998 El Niño. That reef was an experimental design that used electricity. At that time, no one knew how it had protected the reef from the extreme heat. Azeez took the design idea and built an electric reef from steel bars and wired it to a power source on the beach at the Banyan Tree Resort. The small electric current caused the minerals from ocean water to build up on the steel, forming a thick limestone crust that is perfect for coral. The electric reef was situated on the far side of the island, submerged in 15 feet (5 m) of water.

The reef is maintained as a type of underwater topiary. An immense variety of corals grow on it like an underwater garden; all thriving in their sheltered habitat. "Finger corals, hard corals, massive corals. We have tried to plant as many species as we can," Azeez explains. "The reef also attracts both predators and prey, just as a healthy reef ecosystem does." Azeez considers his electric reef as a type of "greenhouse for corals." He believes it will keep a critical mass alive when the next El Niño strikes. Azeez claims, "You can take pieces from the corals on this structure to that one and make your own garden again." This transplant process seems to be working for Azeez in the Maldives. "Our enemy, the real threat," says Azeez, "is global warming."

In a report in National Geographic News, in 1995, it was stated that roughly 10 per cent of the coral colonies in Belize had died. "This coral bleaching is pretty solidly tied to rising ocean temperatures," said Melanie McField, a Belize-based reef scientist with the WWF. "It's a fact that global temperatures have risen. There's lots of data and little argument that increased ocean temperatures are the primary agent of bleaching.

As for tying overall temperature increases to overall global warming, there is still some debate, but less every year. I think the majority of scientists agree that global warming is happening and that it's the root cause of these coral bleaching events. Rather than throwing up our hands and saying 'we can't control that,' we've got to be even more diligent and try even harder to control local impacts such as pollution and overfishing," Mc Field said. Even in view of the unstoppable damage to reefs in the future, conservationists say that given the complex factors affecting coral health, there is still a lot that can be done to help reefs recover if action is taken now.

Protection and Conservation

As the world becomes more aware of the perilous condition of the fragile marine habitat, concern is growing over the future condition and existence of these beautiful areas. As more is learned about what is happening to oceanic ecosystems, more attention is being given to the creation of protected marine areas in order to conserve, manage, and protect ocean resources so that they will not become extinct. The location of conservation areas needs to be carefully chosen to be able to protect and provide for the greatest diversity of species. On June 16, 2006, former president George W. Bush designated the Northwest Hawaiian Islands a national monument.

This created the largest marine protected area in the world, providing important habitat for thousands of marine species that rely on it for survival. Steve McCormick, past president of the Nature Conservancy, "We commend the administration for its foresight and leadership in protecting this incredible area. Designating the northwest Hawaiian Islands as a national monument will ensure that this national treasure will remain healthy and intact for generations to come." This preservation area includes 5,019 square miles (13,000 km2) of coral reefs, comprising 70 per cent of all coral reefs in the United States. The National Geographic News, a consortium of Latin American nations, conservation groups, and United Nations agencies are creating one of the world's largest marine protected areas to be known as the Eastern Tropical Pacific Seascape. The new reserve will cover 521 million acres (211 million ha) of ocean from Costa Rica's Cocos Island to Ecuador's Galápagos Islands. The purpose of the preserve is to protect a wide range of ocean species, expand existing marine reserves, and consolidate current and planned conservation efforts. It will be the largest marine conservation area in the Western Hemisphere. The region already has these currently listed World Heritage Sites: Galápagos Islands and Marine Reserve (in Ecuador), Cocos Island (Costa Rica), and Coiba National Park (Panama). There are several conservation organizations in existence today that are geared towards helping promote healthy ocean ecosystems. Fortunately for the environment, as well as future generations, tens of thousands of people choose to get involved and make a difference.

THE EFFECTS OF GLOBAL WARMING ON OPEN OCEANS

Because the oceans are so vast, much of the predictions made about the Earth's open oceans in light of global warming have been generated by computer models. Because the two most important functions that govern the behaviour of the ocean are temperature and circulation, these calculations are fairly straightforward in computer models and readily calculable. Another thing that makes it possible to model the oceans using computers is that humans have not had as large an impact on the oceans as they have on land.

The Thermohaline Circulation, the Great Conveyor Belt

Because of the interactions of temperature and salinity in the world's oceans, movement is generated that links them all. This linking of the oceans allows heat energy to be transported globally. Cold water is much denser than warm water, so it sinks to the bottom of the ocean while the warmer water rides above it. Water that is high in salt content is also denser, which forces it below water that is lower in salinity. These differences in density are what generate the movement of the oceans. The most important movement is the thermohaline (thermo for "temperature" and haline for "salt content") system, commonly known as the great ocean conveyor belt. This system moves cold dense seawater from the North Atlantic Ocean surface to deep water and then continues through the Indian and Pacific Oceans and back to the Atlantic.

This process takes centuries for an entire cycle to be completed, and it is one of the most important currents in the ocean for regulating heat on the Earth's continents. This is the current that picks up heat in the equatorial region, rises to the ocean's surface, and travels northward via the Gulf Stream/North Atlantic Current past the west coast of Europe, moderating the climate there, making it much milder than it otherwise would be at its northern latitude. As the current reaches its northern limit in the Arctic, it cools and sinks and takes with it the nutrients as well as oxygen and carbon dioxide (CO_2) that it absorbed while it was at the surface of the ocean, creating a sink for CO_2. Deep-sea organisms are able to use the nutrients and oxygen, but the CO_2 does not get used by plants in photosynthesis because the ocean depths are too dark. If global warming intensifies and freshwater from melting glaciers and ice caps is added to the Arctic waters, it will dilute the salinity and slow or stop the vertical mixing between the ocean surface and the deep sea. The result of this would be to slow, or even halt, the conveyor belt.

If this were to happen, warmth would no longer be delivered to Europe via the Gulf Stream, which could put Europe into an ice age. There would be many other effects as well. Diminished vertical mixing between the ocean's surface and depths would reduce the upwelling in temperate and subtropical latitudes. Oxygen would not be effectively transported from the surface to the ocean depths. Over time—centuries— the deep ocean waters would become devoid of oxygen, and they would either become hypoxic (low in oxygen concentrations) or anoxic (completely lacking in oxygen). The dispersal of nutrients from the surface to the depths would also be impaired, leaving deep- sea animals, including commercial fish and squid, without nutrition. This could destroy the fishing industry. As far as the present rate of CO_2 uptake by the oceans, the Pew Climate Centre believes there is disagreement as to whether

CO_2 would be less likely to be absorbed if the ocean depths were not mixing effectively, removing the ocean as a CO_2 sink. This would increase CO_2 concentrations in the atmosphere or it would cause an increase in biological productivity in the upper ocean layers and increase the CO_2 uptake.

The IPCC in their third report, general circulation models (GCMs) predict a wide range of global ocean circulation responses to global warming, ranging from "no response" to a "40 per cent decline in circulation before 2100." Dr. Thomas F. Stocker of the University of Bern's Physics Institute, "changes in the thermohaline circulation are likely with global warming, but the extent of the changes are uncertain at this point." Dr. Peter Robert Gent, chairman of the science steering committee for the Community Climate System Model, working with the National Science Foundation and the U.S. Department of Energy, says he "does not expect a collapse of the thermohaline circulation during the 21st century because of the feedbacks that help stabilize circulation in response to warming."

He warns, however, that "presently, little is known about the feedback mechanisms that affect the thermohaline circulation pattern." Another important consideration, he says, is "that these projections are made under the assumption that atmospheric CO_2 levels do not exceed a doubling of preindustrial levels before 2100. Models that were run basing a CO_2 level above double the preindustrial levels before 2100 did simulate the collapse of the thermohaline circulation by 2100—which could put Europe in an ice age." These are still major unknowns concerning global warming and much more research needs to be done.

Ocean Acidification

Ocean acidification is the ongoing decrease in the pH value of the Earth's oceans caused by their steady uptake of anthropogenic (human-caused) CO_2 from the atmosphere. The German Advisory Council on Global Change, the oceans presently store about 50 times more CO_2 than the atmosphere and 20 times more than the terrestrial biosphere and soils.

The ocean is not only an important CO_2 reservoir, it is also the most important long-term CO_2 sink. Due to pressure differences between the atmosphere and seawater, some of the anthropogenic CO_2 dissolves in the surface layer of the ocean. Over periods of time—decades to centuries to millennia—the CO_2 is carried into the deep sea by ocean currents. Christopher L. Sabine of NOAA's Pacific Marine Environmental Laboratory, the ocean is presently taking up 2 Gt (gigatons) of carbon annually, which just as to the IPCC is the equivalent of roughly 30 per cent of the anthropogenic CO_2 emissions. (A gigaton is one billion tons [900,000,000 metric tons].) The IPCC calculates that between 1800 and 1995, the oceans absorbed about 118 Gt of carbon.

This corresponds to about 48 per cent of the cumulative CO_2 emissions from fossil fuels, or 27–34 per cent of the total anthropogenic CO_2 emissions, including those from land-use changes such as deforestation. Currently, anthropogenic CO_2 can be tracked to a depth of 3,281 feet (1,000 m) in the ocean.

It has not sunk to the bottom of the ocean yet because the vertical mixing process is so slow that it takes a long time. The North Atlantic region is different, however. Because the vertical mixing happens much more readily with the currents there, anthropogenic CO_2 has been traced to a depth of 9,843 feet (3,000 m). In the ocean, CO_2 behaves differently than it does in the atmosphere. In the atmosphere, it is chemically neutral; in the ocean it is chemically active. Dissolved CO_2 contributes to the reduction of the pH value and causes an acidification of seawater, which is a property that can be easily measured. Since the onset of the Industrial Revolution in the 1700s, the pH value of the oceans has dropped (become more acidic) by about 0.11 units. Starting from a slightly alkaline, preindustrial pH value of 8.18, the acidity of the ocean has increased at the surface. Based on modeling done by the IPCC, if atmospheric CO_2 concentrations reach 650 parts per million (ppm) by the year 2100, a decrease in the average pH value by 0.30 units can be expected compared to preindustrial levels. If concentrations increase to 970 ppm, the pH level would drop by 0.46 units. The scale of changes will vary regionally, which will affect the magnitude of any biological effects. Based on the following data from the Royal Society in London, they believe that ocean acidification.

- ... is essentially irreversible during our lifetimes. It will take tens of thousands of years for ocean chemistry to return to a condition similar to that occurring at preindustrial times (about 200 years ago). Our ability to reduce ocean acidification through artificial methods such as through the addition of chemicals is unproven. These techniques will at best be effective only at a very local scale, and could also cause damage to the marine environment. Reducing CO_2 emissions to the atmosphere appears to be the only practical way to minimize the risk of large-scale and long-term changes to the oceans.

All the evidence collected to date points to ocean acidification being caused by human activity—burning fossil fuels, deforestation, and land-use change. The magnitude of acidification can be predicted accurately. What cannot be predicted yet is the physiological effects on various organisms. Headway has been made, however, on discovering the ramifications of acidification on the process of calcification—the process by which organisms such as corals and mollusks make shells and plates from calcium carbonate. The tropical and subtropical corals are expected to be the worst affected. This could destroy coral reef habitats. Phytoplankton and zooplankton, which are major food sources for fish and other animals, will also be negatively affected. This is one area of global warming where research is still in its infancy and needs much more to be done in order to understand both the short- and long-term effects.

Seaweed—A Carbon Sink

Recent studies indicate that seaweed may be capable of sucking CO_2 out of the atmosphere at rates comparable to the world's rain forests. Chung Ik-Kyo, a South Korean environmental scientist, "The ocean's role is neglected because we can't see the vegetation. But under the sea, there is a lot of seaweed and sea grass that can take up carbon dioxide." This concept has such potential, in fact, that 12 Asian-Pacific countries are currently working together to calculate how much CO_2 is being absorbed from the atmosphere by plants and to increase carbon storage through carbon sinks.

In 2007, the United Nations Climate Change Conference in Bali brought together 10,000 participants from 180 countries and adopted the Bali Road Map, various decisions towards reaching a secure climate future. A major issue addressed was identifying those things that can be potentially used as carbon sinks to remove carbon from the air. One possible sink, just as to Chung Ik-Kyo, is seaweed.

There is a tremendous amount of seaweed produced and available each year—8 million tons (7.3 million metric tons)—harvested from wild or cultivated sources.

This theory has met with criticism, however. Some believe it will be a challenge to keep the carbon, once it is absorbed in the seaweed, from reentering the atmosphere at some point. Another negative side effect is the uncertainty about exactly how an increase in seaweed production would affect ocean navigation or fisheries.

The world's current biggest producers of seaweed are China, South Korea, and Japan. Those that favour the plan of using seaweed say that seaweed and algae's rapid rate of photosynthesis (the process of turning CO2 and sunlight into energy and oxygen) make them the perfect candidates for absorbing carbon.

Lee Jae-young of South Korea's fisheries ministry, "Seaweeds can absorb five times more carbon dioxide than plants on land." John Beardall with Australia's Monash University says, "The oceans account for 50 per cent of all the photosynthesis on the Earth.

These are very productive ecosystems." In addition to using seaweed as a CO_2 storage, these same researchers have also suggested that it be used to produce clean-burning biofuels. Skeptics of this concept say that the reason trees are so effective at carbon storage is because they live for many years. Seaweed is grown and harvested in terms of a few months.

They feel that this would make this type of carbon storage difficult to measure and control. Other critics wonder if removing water from the seaweed as it is converted to fuel would require a larger input of energy than that which would be received, making it environmentally unwise. Chung responds by saying, "The idea is still in its infancy. In terms of ball games, we are just in the bullpen, not the main game yet." Currently, South Korea and Japan are the leaders in this type of research.

THE EFFECTS OF GLOBAL WARMING ON COASTAL LOCATIONS

The Pew Centre on Global Climate Change is a non-profit organization that brings together business leaders, policy makers, scientists, and other experts worldwide to create a new approach to managing the problem of global warming-what they refer to as an extremely"controversial issue." Not slanted in any particular way politically or economically, they"approach the issue objectively and base their research and conclusions on sound science, straight talk, and a belief that experts worldwide can work together to protect the climate while sustaining economic growth." The Pew Centre was established in 1998 and has"issued more than 100 reports from top-tier researchers on key climate topics such as economic and environmental impacts and practical domestic and international policy solutions."

The centre's leaders and staff hold briefings with members of Congress and administration officials in the U.S. government, as well as with leaders of international governments. They also work with 43 major corporations in business roles to encourage them to help promote practical solutions to solving the global warming crisis. In January 2007, the Centre was one of the inaugural members of the U.S. Climate Action Partnership-an alliance of major businesses and environmental groups that calls on the federal government to enact legislation requiring significant reductions of greenhouse gas emissions. The Pew Centre on Global Climate Change, in terms of global warming, the biggest impact on estuarine and marine systems will be temperature change, sea-level rise, the availability of water from precipitation and run-off, wind patterns, and storminess. In these oftenfragile systems, temperature has a direct and serious effect. For the sea life living within the ocean, temperature directly affects an organism's biology, such as birth, reproduction, growth, behaviour, and death. The remainder of this part will further explore the specific effects global warming will have on temperature, sea-level rise, wind circulation, aquaculture, algae blooms and disease, and the long-term effects of coastal building.

Temperature

Temperature extremes (both high and low) can be deadly to living organisms. Many species are vulnerable to temperatures just a few degrees higher than what they are accustomed to. Even increases in temperature as little as 1.7°F (1°C) can seriously harm certain species. For example, from 1976-77, the species of reef fish off Los Angeles diminished by 15-25 per cent when temperatures abruptly warmed 1.7°F (1°C). Even if the temperature is not high enough to kill, it can be high enough to negatively influence the organism's life functions such as metabolism, growth, behaviour, and physiological factors such as the timing of reproduction rates of egg and larval development. Temperature also influences where organisms can survive and controls how large a given population can become.

One area where this has long been documented is along the West Coast of North America. The nearshore sea surface is predominantly warm and the offshore area is predominantly cool, creating a region that has long been famous as a rich, diverse fishery. Temperature differences can also influence interaction between species, such as predator-prey, parasite-host, and other relationships that may develop over the struggle for limited resources. If temperatures change the distribution patterns of organisms, it could also change the balance of predators, prey, parasites, and competitors in an ecosystem, completely readjusting balances, food chains, behaviours, and the equilibrium of the ecosystem. Global warming can also change the way that species interact by changing the timing of physiological events. One of the key changes that it could alter is the timing of reproduction for many species. Rising temperature could interfere with the timing of birth being correlated with food availability of that species.

This can be a problem, for example, for bird species that migrate and depend on a specific food source to be available when they reach their breeding grounds. If warmer temperatures have changed the timing and it is a few weeks off of when the food will be available and no longer synchronized with migrating birds, it could leave the birds without available food, threatening their survival. Temperature also plays an important role with oxygen because it directly influences the amount that water can hold. The warmer the water, the less oxygen it holds. As global warming continues to force temperatures to rise, less oxygen will be available for marine species, which could then threaten their survival. A significant amount of data has already been collected on these variables. The Pew Centre, areas that still need much more data collected and research applied concern the ocean temperatures influence on the interactions among organisms, such as predator-prey relationships, parasite-host relationships, and the interplay of species concerning competition for resources.

It is also possible to model the effects of sea-level rise in shallow continental margins, such as flooding wetlands and shoreline erosion. What is more difficult to predict and needs more research are the global warming effects on precipitation amounts, wind patterns, and intense weather. Precipitation directly affects estuaries because it affects the run-off into estuaries, which influences estuarine circulation.

Sea-Level Rise

Sea-level rise will mount from the melting of glacial and polar land ice. The Pew Centre, the effects of sea-level rise will vary by location, how fast the sea level rises, and the biogeochemical responses of the individual ecosystems involved. One of the areas identified as being the most susceptible to damage from sea-level rise is the low-lying, flat wetlands located in the middle and south Atlantic and the Gulf of Mexico (involving states such as Alabama, Florida, Mississippi, Georgia, and Texas). A wetland is an area on shore that has a wet, spongy soil. These areas are also referred to as swamps, marshes, and bogs. As sea levels rise, ocean water will submerge and erode the shorelines. In the natural areas covered with marshes and mangroves (common to Florida and the Gulf states), sea levels will flood the wetlands and waterlog the soils.

Because the plants that live in the wetlands, which do not contain salty water, are not accustomed to salt, the salty ocean water will kill them. Because wetlands provide habitat for wildlife, including several migrating birds, this would also destroy their habitat. Based on research by the Pew Centre, if the wetlands located along the Gulf of Mexico are not bordered by human development (such as homes) and the inland slope is relatively flat, and the sea-level rise is gradual, they would have a chance of surviving by migrating further inland to escape the inundation of salt water. The Pew Centre states that "these areas would be safe at least at the rates of rise currently projected for the next 50 to 100 years." The areas that would be hard hit are those that cannot migrate inland because urbanization has grown right up to their shoreline, effectively removing any potential wetland habitat. This is very detrimental to the environment because wetlands are an important part of the biological productivity of coastal systems. Marshes provide many critical services; they function not only as habitats for wildlife, but as nurseries for breeding and raising young and as refuges from predators. Wetlands function as part of an integrated system.

If they are jeopardized, their loss will affect the availability and transfer of nutrients, the flow of energy, and the availability of natural habitat needed by multitudes of organisms already living there. One of the most unfortunate losses will be those areas where rare, threatened, or endangered plant and animal species live, such as the American alligator, Florida black bear, West Indian manatee, Florida panther, southern bald eagle, snowy egret, and roseate spoonbill. If invading salt water destroys these habitats, all these species could become extinct.

Wind Circulation

Another consideration in the marine environment is wind circulation. Winds are created by the uneven heating at the Earth's surface, specifically the high thermal gradients between the equator and the poles. Under global warming conditions, however, the polar regions will be subjected to higher temperatures, thereby reducing the temperature gradient between the equator and the poles. This could cause a weakening in the overall wind circulation around the Earth. The winds drive the surface currents of the Earth's oceans.

A slowdown of the ocean surface circulation could negatively affect the structure and function of both the open ocean and the near shore ecosystems. Weak currents would make it difficult for any currents to transport nutrients where they need to be deposited. If wind speed and direction are altered, which is what experts at the Pew Centre have predicted will occur with global warming, the productivity of estuarine and marine systems will also be affected.

One of the reasons why some coastal areas are so rich and productive in fishery resources, such as the West Coast of the United States, is because of the coldwater upwelling that occurs along the coastal margins. The thermal properties and circulation patterns function to bring the cold water from the deep ocean waters to the surface along the coastlines, supplying the coastal fisheries with an abundance of fish. Nutrients that are delivered from these deeper waters to the surface help provide for the phytoplankton along the coasts. The upwelling process is controlled by the winds that blow along shore. Upwelling can be minimized, however, if the water column is stratified by differences in temperature or salinity, which act as a barrier to upward movement through the water column. The effects of this type of scenario were played out off the coast of California from 1951 to 1993. The sea surface warmed during this period 2.5°F (1.5°C), resulting in a decreased upwelling of nutrientrich cold waters. This caused a 70 per cent decline in the abundance of zooplankton, which in turn harmed the coastal food webs, negatively affecting fish and birds.

He believes this is possible because most greenhouse gas models project more warming over land than over open oceans, which would increase the temperature contrast between the land and the ocean. This resultant contrast would strengthen the low-pressure cells that usually form over land that are adjacent to offshore high-pressure cells. The relationship of forces between these two pressure cells would create alongshore winds that would create a strong upwelling. On the East Coast of the United States, circulation patterns could slow down transport of nutrients and fish species, such as blue crabs and bluefish, which would lower the abundances of these particular species along the coasts and within estuaries. When their populations are decreased, the coastal ecosystem becomes threatened.

❐

4

Ozone Depletion

Introduction

Ozone is both beneficial and harmful to us. Near the ground, ozone forming as a result of chemical reactions involving traffic pollution and sunlight may cause a number of respiratory problems, particularly for young children, However, high up in the atmosphere (19-30km) in a region known as the stratosphere, ozone filters out incoming radiation from the Sun in the "cell damaging" ultraviolet part of the spectrum. Without ozone in the stratosphere, life on earth would not have evolved. Thus with the development of the ozone layer came the formation of more advanced life forms. Concentrations of ozone in the stratosphere fluctuate naturally in response to variations in weather conditions and amounts of energy being released from the sun, and to major volcanic eruptions. Nevertheless, during the 1970s it was realised that man- made emissions of CFCs and other chemicals used in refrigeration, aerosols and cleansing agents may destroy significant amounts of ozone in the stratosphere, thereby letting through more of the harmful ultraviolet radiation.

Then, in 1985, a large "ozone hole" was discovered the continent of Antarctica during the springtime. This has reappeared every year. In response to this, and additional fears about more widespread global ozone depletion, 24 nations signed the Montreal Protocol on Substances that Deplete the Ozone Layer (1987). This legally binding international treaty called for participating developed nations to reduce the use of CFCs and other ozone depleting substances. In 1990 and again in 1992, subsequent amendments to the Protocol brought forward the phase out date for CFCs for developed countries to 1995. Protecting the ozone layer is essential.

Ultraviolet radiation from the Sun can cause a variety of health problems in humans, including skin cancers, eye cataracts and a reduction in the ability to fight off disease. Furthermore, ultraviolet radiation can be damaging to microscopic life in the surface oceans which forms the basis of the world's food chain, certain varieties of crops including rice and soya, and polymers used in paints and clothing.

A loss of ozone in the stratosphere may even affect the global climate. International agreements have gone a long way to safeguarding this life- supporting shield. Nevertheless, for there to be real and longlasting success, everyone must become part of the solution. Individual efforts taken together can be powerful forces for environmental change. There are a number of things that we, as individuals, can do to both protect the ozone layer.

These include proper disposal of old refrigerators, the use of halon-free fire extinguishers and the recycling of foam and other non-disposable packaging. Finally, we should all be aware that whilst emissions of ozone depleters are now being controlled, the ozone layer is not likely to fully repair itself for several decades. Consequently, we should take precautions when exposing ourselves to the Sun. The Atmospheric Research and Information Centre (aric), through its Atmosphere, Climate and Environment Information Programme, has compiled a series of 20 topical fact sheets concerning the subject of ozone depletion. The series is divided into three parts - the science of ozone depletion, the impacts of ozone depletion, and managing ozone depletion.

Together, they describe what ozone, the ozone layer, ozone depletion and the ozone hole are, how ozone depletion occurs, mankind's influence, its impacts, and the international agreements put in place to control it. The fact sheet series is aimed at students involved in Key Stage 4 of the National Curriculum (GCSE) and higher. Although some of the concepts covered by the fact sheets may be challenging which sometimes contains words and phrases that may seem unfamiliar to the reader. aric hope that the reader will find this fact sheet series a useful information resource on ozone depletion.

WHAT CAUSES OZONE DEPLETION?

It was first suggested, by Drs. M. Molina and S. Rowland in 1974, that a man-made group of compounds known as the chloro-fluorocarbons (CFCs) were likely to be the main source of ozone depletion. However, this idea was not taken seriously until the discovery of the ozone hole over Antarctica in 1985. Chlorofluorocarbons are not "washed" back to Earth by rain or destroyed in reactions with other chemicals. They simply do not break down in the lower atmosphere and they can remain in the atmosphere from 20 to 120 years or more. As a consequence of their relative stability, CFCs are instead transported into the stratosphere where they are eventually broken down by ultraviolet radiation, releasing free chlorine.

The chlorine becomes actively involved in the process of destruction of ozone. The net result is that two molecules of ozone are replaced by three of molecular oxygen, leaving the chlorine free to repeat the process:

$$Cl + O_3 \rightarrow ClO + O_2$$
$$ClO + O \rightarrow Cl + O_2$$

Ozone is converted to oxygen, leaving the chlorine atom free to repeat the process up to 100,000 times, resulting in a reduced level of ozone. Bromine compounds, or halons, can also destroy stratospheric ozone. Compounds containing chlorine and bromine from manmade synthetic compounds are known as industrial halocarbons.

How Long has Ozone Depletion been Occurring?

Based on data collected since the 1950s, scientists have determined that ozone levels were relatively stable until the late 1970s. Severe depletion over the Antarctic has been occurring since 1979 and a general downturn in global ozone levels has been observed since the early 1980s.

How much of the Ozone Layer has been depleted around the world?

Global ozone levels have declined an average of about 3 per cent between 1979 and 1991. This rate of decline is about three times faster than that recorded in the 1970s. In addition to Antarctica, ozone depletion now affects almost all of North America, Europe, Russia, Australia, New Zealand, and a sizeable part of South America. Short term losses of ozone can be much greater than the long term average. In Canada, ozone depletion is usually greatest in the late winter and early spring. In 1993, for example, average ozone values over Canada were 14 per cent below normal from January to April.

ORIGIN OF OZONE

The photochemical mechanisms that give rise to the ozone layer were discovered by the British physicist Sidney Chapman in 1930. Ozone in the Earth's stratosphere is created by ultraviolet light striking oxygen molecules containing two oxygen atoms (O_2), splitting them into individual oxygen atoms (atomic oxygen); the atomic oxygen then combines with unbroken O_2 to create ozone, O_3. The ozone molecule is also unstable (although, in the stratosphere, long-lived) and when ultraviolet light hits ozone it splits into a molecule of O_2 and an atom of atomic oxygen, a continuing process called the ozone-oxygen cycle, thus creating an ozone layer in the stratosphere, the region from about 10 to 50 km (32,000 to 164,000 feet) above Earth's surface. About 90 per cent of the ozone in our atmosphere is contained in the stratosphere. Ozone concentrations are greatest between about 20 and 40 km, where they range from about 2 to 8 parts per million. If all of the ozone were compressed to the pressure of the air at sea level, it would be only a few millimeters thick.

Ultraviolet Light and Ozone

Although the concentration of the ozone in the ozone layer is very small, it is vitally important to life because it absorbs biologically harmful ultraviolet (UV) radiation coming from the Sun. UV radiation is divided into three categories, based on its wavelength; these are referred to as UV-A (400–315 nm), UV-B (315–280 nm), and UV-C (280–100 nm). UV-C, which would be very harmful to all living things, is entirely screened out by ozone at around 35 km altitude. UV-B radiation can be harmful to the skin and is the main cause of sunburn; excessive exposure can also cause genetic damage, resulting in problems such as skin cancer. The ozone layer is very effective at screening out UV-B; for radiation with a wavelength of 290 nm, the intensity at the top of the atmosphere is 350 million times stronger than at the Earth's surface. Nevertheless, some UV-B reaches the surface. Most UV-A reaches the surface; this radiation is significantly less harmful, although it can potentially cause genetic damage.

Distribution of Ozone in the Stratosphere

The thickness of the ozone layer—that is, the total amount of ozone in a column overhead—varies by a large factor worldwide, being in general smaller near the equator and larger towards the poles. It also varies with season, being in general thicker during the spring and thinner during the autumn in the northern hemisphere. The reasons for this latitude and seasonal dependence are complicated, involving atmospheric circulation patterns as well as solar intensity.

Since stratospheric ozone is produced by solar UV radiation, one might expect to find the highest ozone levels over the tropics and the lowest over polar regions. The same argument would lead one to expect the highest ozone levels in the summer and the lowest in the winter. The observed behaviour is very different: most of the ozone is found in the mid-to-high latitudes of the northern and southern hemispheres, and the highest levels are found in the spring, not summer, and the lowest in the autumn, not winter in the northern hemisphere. During winter, the ozone layer actually increases in depth. This puzzle is explained by the prevailing stratospheric wind patterns, known as the Brewer-Dobson circulation. While most of the ozone is indeed created over the tropics, the stratospheric circulation then transports it poleward and downward to the lower stratosphere of the high latitudes. However in the southern hemisphere, owing to the ozone hole phenomenon, the lowest amounts of column ozone found anywhere in the world are over the Antarctic in the southern spring period of September and October.

The ozone layer is higher in altitude in the tropics, and lower in altitude in the extratropics, especially in the polar regions. This altitude variation of ozone results from the slow circulation that lifts the ozone-poor air out of the troposphere into the stratosphere. As this air slowly rises in the tropics, ozone is produced by the overhead sun which photolyzes oxygen molecules. As this slow circulation bends towards the mid-latitudes, it carries the ozone-rich air from the tropical middle stratosphere to the mid-and-high latitudes lower stratosphere. The high ozone concentrations at high latitudes are due to the accumulation of ozone at lower altitudes.

The Brewer-Dobson circulation moves very slowly. The time needed to lift an air parcel from the tropical tropopause near 16 km (50,000 ft) to 20 km is about 4–5 months (about 30 feet (9.1 m) per day). Even though ozone in the lower tropical stratosphere is produced at a very slow rate, the lifting circulation is so slow that ozone can build up to relatively high levels by the time it reaches 26 km.

Ozone amounts over the continental United States (25°N to 49°N) are highest in the northern spring (April and May). These ozone amounts fall over the course of the summer to their lowest amounts in October, and then rise again over the course of the winter. Again, wind transport of ozone is principally responsible for the seasonal evolution of these higher latitude ozone patterns. The total column amount of ozone generally increases as we move from the tropics to higher latitudes in both hemispheres. However, the overall column amounts are greater in the northern hemisphere high latitudes than in the southern hemisphere high latitudes. In addition, while the highest amounts of column ozone over the Arctic occur in the northern spring (March–April), the opposite is true over the Antarctic, where the lowest amounts of column ozone occur in the southern spring (September–October).

Ozone Depletion

The ozone layer can be depleted by free radical catalysts, including nitric oxide (NO), nitrous oxide (N_2O), hydroxyl (OH), atomic chlorine (Cl), and atomic bromine (Br). While there are natural sources for all of these species, the concentrations of chlorine and bromine have increased markedly in recent years due to the release of large quantities of man-made organohalogen compounds, especially chlorof-luorocarbons (CFCs) and bromofluorocarbons. These highly stable compounds are capable of surviving the rise to the stratosphere, where Cl and Br radicals are liberated by the action of ultraviolet light. Each radical is then free to initiate and catalyze a chain reaction capable of breaking down over 100,000 ozone molecules. The breakdown of ozone in the stratosphere results in the ozone molecules being unable to absorb ultraviolet radiation. Consequently, unabsorbed and dangerous ultraviolet-B radiation is able to reach the Earth's surface. Ozone levels, over the northern hemisphere, have been dropping by 4 per cent per decade. Over approximately 5 per cent of the Earth's surface, around the north and south poles, much larger (but seasonal) declines have been seen; these are the ozone holes. In 2009, nitrous oxide (N_2O) was the largest ozone-depleting substance emitted through human activities.

Regulation

In 1978, the United States, Canada and Norway enacted bans on CFC-containing aerosol sprays that are thought to damage the ozone layer. The European Community rejected an analogous proposal to do the same. In the U.S., chlorofluorocarbons continued to be used in other applications, such as refrigeration and industrial cleaning, until after the discovery of the Antarctic ozone hole in 1985. After negotiation of an international treaty (the Montreal Protocol), CFC production was sharply limited beginning in 1987 and phased out completely by 1996.

On August 2, 2003, scientists announced that the depletion of the ozone layer may be slowing down due to the international ban on CFCs. Three satellites and three ground stations confirmed that the upper atmosphere ozone depletion rate has slowed down significantly during the past decade. The study was organized by the American Geophysical Union. Some breakdown can be expected to continue due to CFCs used by nations which have not banned them, and due to gases which are already in the stratosphere. CFCs have very long atmospheric lifetimes, ranging from 50 to over 100 years, so the final recovery of the ozone layer is expected to require several lifetimes.

Compounds containing C–H bonds have been designed to replace the function of CFC's (such as HCFC), since these compounds are more reactive and less likely to survive long enough in the atmosphere to reach the stratosphere where they could affect the ozone layer. While being less damaging than CFC's, HCFC's also have a significant negative impact on the ozone layer. HCFC's are also being phased out.

Why is the Ozone Layer Important?

Ozone's unique physical properties allow the ozone layer to act as our planet's sunscreen, providing an invisible filter to help protect all life forms from the Sun's damaging ultraviolet (UV) rays. Most incoming UV radiation is absorbed by ozone and prevented from reaching the Earth's surface. Without the protective effect of ozone, life on Earth would not have evolved in the way it has.

OXYGEN AND OZONE

Ozone is a form of oxygen. The oxygen we breathe is in the form of oxygen molecules (O_2) - two atoms of oxygen bound together. Ozone, on the other hand, consists of three atoms of oxygen bound together (O_3). Most of the atmosphere's ozone occurs in the upper atmosphere called the stratosphere.

Ozone is colourless and has a very harsh odour. Normal oxygen, which we breathe, has two oxygen atoms and is colourless and odourless. Ozone is much less common than normal oxygen. Out of each 10 million air molecules, about 2 million are normal oxygen, but only 3 are ozone.

Where is Ozone Found?

Most ozone is produced naturally in the upper atmosphere or stratosphere. While ozone can be found through the entire atmosphere, the greatest concentration occurs at an altitude of about 25 km. This band of ozone-rich air is known as the "ozone layer". The stratosphere ranges from an altitude of 10km to 50km and it lies above the troposphere (or lower atmosphere).

Stratospheric ozone shields us from the harmful effects of the Sun's ultraviolet radiation. Ozone also occurs in very small amounts in the troposphere (200 parts per billion). It is produced at ground level through a reaction between sunlight and volatile organic compounds (VOCs) and nitrogen oxides (NOx), some of which are produced by human activities such as driving cars.

Ground-level ozone is a component of urban smog, and can damage certain crops and trees, and can be harmful to human health. As there is increased solar radiation during the summer months, there tends to be a larger number of urban ozone episodes at this time. Ground level ozone is blown easily by surface winds to regions well beyond its source. As a pollutant it should not be confused with the separate problem of stratospheric ozone depletion.

Even though both types of ozone are exactly the same molecule, their presence in different parts of the atmosphere has very different consequences. Stratospheric ozone blocks harmful solar radiation - all life on Earth has adapted to this filtered solar radiation. Groundlevel ozone, in contrast, is simply a pollutant. It will absorb some incoming solar radiation, but it cannot make up for stratospheric ozone loss. This fact sheet series is concerned with stratospheric ozone depletion.

OTHER EFFECTS OF OZONE DEPLETION

As well as the effects to human health, land plants and aquatic life, which may occur as a consequence of ozone depletion, there are other impacts which could result from prolonged destruction of ozone in the stratosphere. These include damage to polymers used in buildings, paints and packaging, and changes in biogeochemical cycles affecting ground-level pollution (smog), acid rain and even climate change.

Damage to Polymers

Ozone depletion will cause many materials to degrade faster. These materials include PVC (used in window and door frames, pipes and gutters, etc.), nylon and polyester. They are all composed of compounds known as polymers. Synthetic polymers, naturally occurring biopolymers, as well as some other materials of commercial interest are adversely affected by solar UV radiation. Today's materials are somewhat protected from UV-B by special additives. Therefore, any increase in solar UV-B levels as a result of ozone depletion will therefore accelerate their breakdown, limiting how long they are useful outdoors. Shorter wavelength (i.e. more energetic) UV-B radiation is mainly responsible for photo-damage ranging from discolouration to loss of mechanical integrity in polymers exposed to solar radiation. The use of higher levels of conventional light stabilisers in polymerbased materials are likely to be employed to mitigate the effects of increased UV levels in sunlight. However, it is not certain how resistant such light stabilisers are themselves to increased levels of UV-radiation. In addition, their use will add to the cost of plastic products in target applications. With plastics rapidly displacing conventional materials in numerous applications, this is an important consideration particularly in the developing world. It is not certain yet how other materials, including rubber, paints, wood, document and textiles will be affected by increased UV radiation resulting from ozone depletion.

Effects on Biogeochemical Cycles

Increases in solar UV radiation could affect terrestrial and aquatic biogeochemical cycles, thereby altering both sources and sinks of greenhouse and chemically-important trace gases e.g., carbon dioxide (CO_2), carbon monoxide (CO), carbonyl sulphide (COS) and possibly other gases, including ozone. These potential changes would contribute to biosphere-atmosphere feedbacks that attenuate or reinforce the atmospheric build-up of these gases. Likely effects include an increase in smog in urban centres, and acid rain in rural areas.

Effects on Climate

Whilst increases of UV radiation as a result of ozone depletion may affect the production and removal of carbon dioxide, the main greenhouse gas, ozone depletion itself can influence the global climate. Ozone is also a greenhouse gas, and as well as filtering out the incoming short-wave solar radiation, can absorb much of the outgoing long-wave terrestrial radiation (infrared radiation). If stratospheric ozone is destroyed, ozone's greenhouse effect is reduced and this could lead to a global cooling, offsetting some of the warming that may be occurring as a result of man-made emissions of carbon dioxide, methane and nitrous oxide. Ironically, when the ozone layer starts to repair itself in the next century as a result of a control on CFCs, this cooling potential will be lost. More significantly, the replacement chemicals to CFCs, the HCFCs, which themselves do not harm the ozone layer, are very strong greenhouse gases, and are further contributing to the potential problem of global warming.

Montreal Protocol

In 1985 an international agreement, the Vienna Convention, was signed after three years of negotiating under the auspices of the United Nations Environment Programme. The Vienna Convention established mechanisms for international co-operation in research, monitoring, and exchange of data on emissions, on concentrations of CFCs and halons, and on the status of stratospheric ozone. It also set a framework for international negotiations on actual reductions of emissions. That same year, 1985, marked another seminal development in the evolution of scientific and public policy recognition of the stratospheric ozone issue - the discovery of the Antarctic ozone hole. On the basis of the Vienna Convention (1985), the Montreal Protocol on Substances that Deplete the Ozone Layer was negotiated and signed by 24 countries and by the European Economic Community in September 1987. The Protocol called for the Parties to phase down the use of CFCs, halons and other man-made halocarbons.

The Protocol

The Montreal Protocol represented a landmark in the international environmentalist movement. For the first time whole countries were legally bound to reducing and eventually phasing out altogether the use of CFCs and other ozone depleting chemicals. Failure to comply was accompanied by stiff penalties. The original Protocol aimed to decrease the use of chemical compounds destructive to ozone in the upper atmosphere by 50 per cent by the year 1999. The agreement was supplemented by agreements made in London in 1990 and in Copenhagen in 1992, by which the same countries promised to stop using CFCs and most of the other chemical compounds destructive to ozone by the end of 1995. The Protocol has been subsequently amended twice more, at Montreal in 1997 and at Beijing in 1999. In most cases it has been fairly easy to develop and introduce compounds and methods to replace CFC compounds.

CFC use in aerosols and foam plastic packaging has already been abandoned in most countries. On the other hand, compounds capable of replacing CFC compounds in cooling devices and insulating materials are still under development. In order to deal with the special difficulties experienced by developing countries it was agreed that they would be given 10 years grace, so long as their use of CFCs did not grow significantly. China and India, for example, are strongly increasing the use of air conditioning and cooling devices. Using CFC compounds in these devices would be cheaper than using replacement compounds harmless to ozone. An international fund has therefore been set up to help these countries to introduce new and environmentally more friendly technologies and chemicals. The depletion of the ozone layer is a world-wide problem which does not respect the frontiers between different countries. It can only be affected through determined international co-operation.

Can Ozone Depletion be Reserved

Can the hole in the ozone layer be repaired? Yes. If concentrations of ozone-destroying chemicals are reduced, the natural balance between ozone creation and destruction can be restored. However, this might require the complete elimination of CFCs, halons, carbon tetrachloride, methyl chloroform, and methyl bromide. In late 1991, scientists estimated that even with the current global schedule to eliminate ozone-destroying substances, the ozone layer would not return to 'normal' (pre-1980 chlorine levels) until the middle of the 21st century. Nevertheless, the 1998 World Meteorological Organiszation Scientific Assesment of Ozone Depletion observed that abundance of ozone- depleting compounds in the lower atmosphere is now slowly declining from a peak in 1994.

Phasing Out CFCS

The initial concern about the ozone layer in the 1970s led to a ban on the use of CFCs as aerosol propellants in several countries, including the U.S. However, production of CFCs and other ozone-depleting substances grew rapidly afterwards as new uses were discovered. Through the 1980s, other uses expanded and the world's nations became increasingly concerned that these chemicals would further harm the ozone layer. In 1985, the Vienna Convention was adopted to formalise international co-operation on this issue. Additional efforts resulted in the signing of the Montreal Protocol in 1987. After the original Protocol was signed, new measurements showed worse damage to the ozone layer than was originally expected.

In 1992, reacting to the latest scientific assessment of the ozone layer, the Parties decided to completely end production of halons by the beginning of 1994 and of CFCs by the beginning of 1996 in developed countries. Between 1986 and 1991, world-wide consumption of CFCs-11, CFC- 12 and CFC-113 decreased by 40 per cent, ahead of the schedule outlined in the Montreal Protocol and faster even than called for in the more ambitious 1990 London Amendment to the Montreal Protocol. Manufacturers, who were earlier convinced that CFCs were unique and irreplaceable, were finding themselves moving quickly to alternative processes and chemicals. Hydrocarbons have replaced CFCs as aerosol propellants and as blowing agents for foams, and as cleansing solvents in electronics manufacturing. HCFCs, which have much smaller ozone depleting potentials have replaced CFCs for refrigeration and air conditioning.

Are International Agreements Enough

Without the Montreal Protocol, continued use of CFCs and other compounds would have tripled the stratospheric abundance of chlorine and bromine by 2050. Because of measures taken under the Protocol, emissions of ozone-depleting substances are already falling. Under current agreements, the stratospheric concentrations of chlorine and bromine are expected to reach their maximum within a few years and then slowly decline, although concentrations of chlorine are already falling in the troposphere. With evidence that international agreements to phase out the use of ozone depleting chemicals appear to be working, both NASA (the North American Space Agency) and NOAA (the National Oceanographic and Atmospheric Administration) in the United States of America have expressed confidence that, all other things being equal, the stratospheric ozone layer should return to normal by the middle of the next century.

The recovery of the ozone layer will be gradual because of the long times required for CFCs to be removed from the atmosphere; some take as long as several hundred years. Nevertheless, the likelihood remains that deep ozone holes will continue to form annually in the polar regions, well into the next century. This situation will persist until stratospheric chlorine levels decrease. The Intergovernmental Panel on Climate Change (IPCC), even if the control measures of the 1990 London and 1992 Copenhagen Amendments were to be implemented by all nations, the abundance of stratospheric chlorine and bromine will increase over the next several years. The Antarctic ozone hole, caused by industrial halocarbons, will therefore recur each spring. In addition, since these gases are also responsible for the observed reduction in middle and high latitude stratospheric ozone, the depletion at these latitudes is predicted to continue unabated for at least 5 to 10 years.

OBSERVED CHANGES IN STRATOSPHERIC OZONE

Global and hemispheric-scale variations in stratospheric ozone can be quantified from extensive observational records covering the past 20 to 30 years. There are numerous ways to measure ozone in the atmosphere, but they fall broadly into two categories: measurements of column ozone and measurements of the vertical profile of ozone. Approximately 90 per cent of the vertically integrated ozone column resides in the stratosphere. There are more independent measurements, longer time-series, and better global coverage for column ozone. Regular measurements of column ozone are available from a network of surface stations, mostly in the mid-latitude NH, with reasonable coverage extending back to the 1960s. Near-global, continuous column ozone data are available from satellite measurements beginning in 1979. The different observational data sets can be used to estimate past ozone changes, and the differences between data sets provide a lower bound of overall uncertainty.

The differences indicate good overall agreement between different data sources for changes in column ozone, and thus we have reasonable confidence in describing the spatial and temporal characteristics of past changes. Five data sets of zonal and monthly mean column ozone values developed by different scientific teams were used to quantify past ozone changes; they include ground-based measurements covering 1964–2001, and several different satellite data sets extending in time over 1979–2001. The analyses first remove the seasonal cycle from each data set and the deviations are area weighted and expressed as anomalies with respect to the period 1964–1980. The global ozone amount shows decreasing values between the late 1970s and the early 1990s, a relative minimum during 1992–1994, and slightly increasing values during the late 1990s.

Global ozone for the period 1997–2004 was approximately 3 per cent below the 1964–1980 average values. Since systematic global observations began in the mid-1960s, the lowest annually averaged global ozone occurred during 1992–1993. These changes are evident in each of the available global data sets. No significant long-term changes in column ozone have been observed in the tropics. Column ozone changes averaged over mid-latitudes are significantly larger in the SH than in the NH; averaged for the period 1997–2001, SH values are 6 per cent below pre-1980 values, whereas NH values are 3 per cent lower. Also, there is significant seasonality to the NH mid-latitude losses whereas long-term SH losses are about 6 per cent year round. The most dramatic changes in ozone have occurred during the spring season over Antarctica, with the development during the 1980s of a phenomenon known as the ozone hole.

The ozone hole now recurs every spring, with some interannual variability and occasional extreme behaviour. In most years the ozone concentration is reduced to nearly zero over a layer several kilometres deep within the lower stratosphere in the Antarctic polar vortex. Since the early 1990s, the average October column ozone poleward of 63°S has been more than 100 DU below pre-ozone-hole values with up to a 70 per cent local decrease for periods of a week or so. Compared with the Antarctic, Arctic ozone abundance in the winter and spring is highly variable because of interannual variability in chemical loss and in dynamical transport. Dynamical variability within the winter stratosphere leads to changes in ozone transport to high latitudes, and these transport changes are correlated with polar temperature variability–with less ozone transport being associated with lower temperatures. Low temperatures favour halogen-induced chemical ozone loss. Thus, in recent decades, halogen-induced polar ozone chemistry has acted in concert with the dynamically induced ozone variability, and has led to Arctic column ozone losses of up to 30 per cent in particularly cold winters. In particularly dynamically active, warm winters, however, the estimated chemical ozone loss has been very small. Changes in the vertical profile of ozone are derived primarily from satellites, ground-based measurements and balloon-borne ozonesondes. Long records from ground-based and balloon data are available mainly for stations over NH mid-latitudes. Ozone profile changes over NH mid-latitudes exhibit two maxima, with decreasing trends in the upper stratosphere and in the lower stratosphere during the period 1979–2000. Ozone profile trends show a minimum near 30 km. The vertically integrated profile trends are in agreement with the measured changes in column ozone.

Observed Changes in ODSS

As a result of reduced emissions because of the Montreal Protocol and its Amendments and Adjustments, mixing ratios for most ODSs have stopped increasing near the Earth's surface. The response to the Protocol, however, is reflected in quite different observed behaviour for different substances. By 2003, the mixing ratios for CFC-12 were close to their peak, CFC-11 had clearly decreased, while methyl chloroform had dropped by 80 per cent from its maximum. Halons and HCFCs are among the few ODSs whose mixing ratios were still increasing in 2000. Halons contain bromine, which is on average 40 to 50 times more efficient on a per-atom basis at destroying stratospheric ozone than chlorine.

However, growth rates for most halons have steadily decreased during recent years. Furthermore, increases in tropospheric bromine from halons have been offset by the decline observed for methyl bromide since 1998. Atmospheric amounts of HCFCs continue to increase because of their use as CFC substitutes. Chlorine from HCFCs has increased at a fairly constant rate of 10 ppt Cl yr–1 since 1996, although HCFCs accounted for only 5 per cent of all chlorine from long-lived gases in the atmosphere by 2000. The ODPs of the most abundant HCFCs are only about 5–10 per cent of those of the CFCs. The behaviour of CFC-11 is representative of the behaviour of CFCs in general, and the behaviour of HCFC-22 and HFC-134a is representative of HCFCs in general. The most rapid growth rates of CFC-11 occurred in the 1970s and 1980s. The largest emissions were in the NH, and concentrations in the SH lagged behind those in the NH, consistent with an inter-hemispheric mixing time scale of about 1 to 2 years.

In recent years, following the implementation of the Montreal Protocol, the observed growth rate has declined, concentrations appear to be at their peak and, as emissions have declined, the inter-hemispheric gradient has almost disappeared. In contrast, HCFC-22 and HFC-134a concentrations are still growing rapidly and there is a marked inter-hemispheric gradient. Ground-based observations suggest that by 2003 the cumulative totals of both chlorine- and bromine-containing gases regulated by the Montreal Protocol were decreasing in the lower atmosphere. Although tropospheric chlorine levels peaked in the early 1990s and have since declined, atmospheric bromine began decreasing in 1998. The net effect of changes in the abundance of both chlorine and bromine on the total ozone-depleting halogens in the stratosphere is estimated roughly by calculating the equivalent effective stratospheric chlorine. The calculation of EESC includes consideration of the total amount of chlorine and bromine accounted for by long-lived halocarbons, how rapidly these halocarbons degrade and release their halogen in the stratosphere and a nominal lag time of three years to allow for transport from the troposphere into the stratosphere. The tropospheric observational data suggest that EESC peaked in the mid-1990s and has been decreasing at a mean rate of 22 ppt yr–1 over the past eight years. Direct stratospheric measurements show that stratospheric chlorine reached a broad plateau after 1996, characterized by variability.

Observed Changes in Stratospheric Aerosols, Water Vapour, Methane and Nitrous Oxide

In addition to ODSs, stratospheric ozone is influenced by the abundance of stratospheric aerosols, water vapour, methane and nitrous oxide. During the past three decades, aerosol loading in the stratosphere has primarily reflected the effects of a few volcanic eruptions that inject aerosol and its gaseous precursors into the stratosphere. The most noteworthy of these eruptions are El Chichón and Pinatubo. The 1991 Pinatubo eruption likely had the largest impact of any event in the 20th century, producing about 30

Tg of aerosol that persisted into at least the late 1990s. Current aerosol loading, which is at the lowest observed levels, is less than 0.5 Tg, so the Pinatubo event represents nearly a factor of 100 enhancement relative to non-volcanic levels. The source of the non-volcanic stratospheric aerosol is primarily carbonyl sulfide and there is general agreement between the aerosols estimated by modelling the transformation of observed OCS to sulphate aerosols and observed aerosols.

However, there is a significant dearth of SO_2 measurements, and the role of tropospheric SO_2 in the stratospheric aerosol budget–while significant– remains a matter of some uncertainty. Because of the high variability of stratospheric aerosol loading it is difficult to detect trends in the non-volcanic aerosol component. Trends derived from the late 1970s to the current period are likely to encompass a value of zero. The recent Stratospheric Processes And their Role in Climate Assessment of Upper Tropospheric and Stratospheric Water Vapour provided an extensive review of data sources and quality for stratospheric water vapour, together with detailed analyses of observed seasonal and interannual variability. The longest continuous reliable data set is at a single location is based on balloon-borne frost-point hygrometer measurements and dates back to 1980. Over the period 1980–2000, a statistically significant positive trend of approximately 1 per cent yr–1 is observed at all levels between about 15 and 26 km in altitude. However, although a linear trend can be fitted to these data, there is a high degree of variability in the infrequent sampling, and the increases are neither continuous nor steady. Long- term increases in stratospheric water vapour are also inferred from a number of other ground-based, balloon, aircraft and satellite data sets spanning approximately 1980–2000, although the time records are short and the sampling uncertainty is high in many cases.

Global stratospheric water vapour measurements have been made by the Halogen Occultation Experiment satellite instrument for more than a decade. Interannual changes in water vapour derived from HALOE data show excellent agreement with Polar Ozone and Aerosol Measurement satellite data, and also exhibit strong coherence with tropical tropopause temperature changes. The Boulder and HALOE data show reasonable agreement for the early part of the record but there is an offset after 1997, with the Boulder data showing higher values than HALOE measurements. As a result, linear trends derived from these two data sets for the period 1992–2004 show very different results. The reason for the differences between the balloon and satellite data is unclear at present, but the discrepancy calls into question interpretation of water vapour trends derived from short or infrequently sampled data records.

It will be important to reconcile these differences because these data sets are the two longest and most continuous data records available for stratospheric water vapour. It is a challenge to explain the magnitude of the water vapour increases seen in the Boulder frost-point data. Somewhat less than half the observed increase of about 10 per cent per decade can be explained as a result of increasing tropospheric methane. The remaining increase could be reconciled with a warming of the tropical tropopause of approximately 1 K per decade, assuming that air entered the stratosphere with water vapour in equilibrium with ice. However, observations suggest that the tropical tropopause has cooled slightly for this time period, by approximately 0.5 K per decade, and risen slightly in altitude by about

20 m per decade. Although regional-scale processes may also influence stratospheric water vapour, there is no evidence for increases in tropopause temperature in these regions either. From this perspective, the extent of the decadal water vapour increases inferred from the Boulder measurements is inconsistent with the observed tropical tropopause cooling, and this inconsistency limits confidence in predicting the future evolution of stratospheric water vapour. The atmospheric abundance of methane has increased by a factor of about 2.5 since the pre-industrial era. Measurements of methane from a global monitoring network showed increasing values through the 1990s, but approximately constant values during 1999–2002. Changes in stratospheric methane have been monitored on a global scale using HALOE satellite measurements since 1991. The HALOE data show increases in lower stratospheric methane during 1992–1997 that are in reasonable agreement with tropospheric increases during this time.

In the upper stratosphere the HALOE data show an overall decrease in methane since 1991, which is likely attributable to a combination of chemical and dynamical influences. Measurements of tropospheric N_2O show a consistent increase of about 3 per cent per decade. Because tropospheric air is transported into the stratosphere, these positive N_2O trends produce increases in stratospheric reactive nitrogen which plays a key role in ozone photochemistry. Measurements of stratospheric column NO_2 from the SH and the NH mid-latitudes show long-term increases of approximately 6 per cent per decade and these are reconciled with the N_2O changes by considering effects of changing levels of stratospheric ozone, water vapour and halogens.

Observed Temperature Changes in the Stratosphere

There is strong evidence of a large and significant cooling in most of the stratosphere since 1980. Recent updates to the observed changes in stratospheric temperature and to the understanding of those changes have been presented of WMO. Current long-term monitoring of stratospheric temperature relies on satellite instruments and radiosonde analyses. The Microwave Sounding Unit and Stratospheric Sounding Unit instruments record temperatures in several 10–15 km thick layers between 17 and 50 km in altitude. Radiosonde trend analyses are available up to altitudes of roughly 25 km. Determining accurate trends with these data sets is difficult. In particular the radiosonde coverage is not global and suffers from data quality concerns, whereas the satellite trend data is a result of merging data sets from several different instruments. It reveals a strong imprint of 1 to 2 years of warming following the volcanic eruptions of Agung, El Chichón and Mt. Pinatubo. When these years are excluded from long-term trend analyses, a significant global cooling is seen in both the radiosonde and the satellite records over the last few decades. This cooling is significant at all levels of the stratosphere except the 30 km level in the SSU record. The largest global cooling is found in the upper stratosphere, where it is fairly uniform in time at a rate of about 2 K per decade. In the lower stratosphere this long-term global cooling manifests itself as more of a step-like change following the volcanic warming events. The cooling also varies with latitude.

The lower stratosphere extratropics show a cooling of 0.4–0.8 K per decade in both hemispheres, which remains roughly constant throughout the year. At high latitudes most of the cooling, up to 2 K per decade for both hemispheres, occurs during spring. Much recent progress has been made in modelling these temperature trends. Models range from one-dimensional fixed dynamical heating rate calculations to three-dimensional coupled chemistry-climate models; many of their findings are presented later in this stage. FDH is a simple way of determining the radiative response to an imposed change whilst 'fixing' the background dynamical heating to its climatological value; this makes the calculation of temperature change simpler, as only a radiation model needs to be used. Changes in dynamical circulations simulated by models can lead to effects over latitudinal bands, which vary between models.

However, dynamics cannot easily produce a global mean temperature change. Because of this, global mean temperature is radiatively controlled and provides an important focus for attribution. For the global mean in the upper stratosphere the models suggest roughly equal contributions to the cooling from ozone decreases and carbon dioxide increases. In the global mean mid-stratosphere there appears to be some discrepancy between the SSU and modelled trends near 30 km: models predict a definite radiative cooling from carbon dioxide at these altitudes, which is not evident in the SSU record. In the lower stratosphere the cooling from carbon dioxide is much smaller than higher in the stratosphere. Although ozone depletion probably accounts for up to half of the observed cooling trend, it appears that a significant cooling from another mechanism may be needed to account for the rest of the observed cooling.

One possible cause of this extra cooling could be stratospheric water vapour increases. However, stratospheric water vapour changes are currently too uncertain to pinpoint their precise role. Tropospheric ozone increases have probably contributed slightly to lower stratospheric cooling by reducing upwelling thermal radiation; one estimate suggests a cooling of 0.05 K per decade at 50 hPa and a total cooling of up to 0.5 K over the last century. The springtime cooling in the Antarctic lower stratosphere is almost certainly nearly all caused by stratospheric ozone depletion. However, the similar magnitude of cooling in the Arctic spring does not seem to be solely caused by ozone changes, which are much smaller than in the Antarctic; interannual variability may contribute substantially to the cooling observed in the Arctic wintertime. One mechanism for altering temperatures in the upper troposphere and lower stratosphere comes from the direct radiative effects of halocarbons. In contrast with the role of carbon dioxide, halocarbons can actually warm the upper tropospheric and lower stratospheric region.

In summary, stratospheric temperature changes over the past few decades are significant and there are clear quantifiable features of the contributions from ozone, carbon dioxide and volcanism in the past stratospheric temperature record. More definite attribution of the causes of these trends is limited by the short timeseries. Future increases of carbon dioxide can be expected to substantially cool the upper stratosphere.

However, this cooling could be partially offset by any future ozone increase. In the lower stratosphere, both ozone recovery and halocarbon increases would warm this region compared with the present. Any changes in stratospheric water vapour or changes in tropospheric conditions, such as high-cloud properties and tropospheric ozone, would also affect future temperatures in the lower stratosphere. Furthermore, circulation changes can affect temperatures over sub-global scales, especially at midand polar latitudes. Several studies have modelled parts of these expected temperature changes.

Ozone in the Atmosphere and its Role in Climate

A number of important concepts concerning the ozone layer and its role in the stratosphere. Ozone, like water vapour and carbon dioxide, is an important and naturally occurring greenhouse gas; that is, it absorbs and emits radiation in the thermal infrared, trapping heat to warm the Earth's surface. In contrast to the so-called wellmixed greenhouse gases, stratospheric ozone has two distinguishing properties. First, its relatively short chemical lifetime means that it is not uniformly mixed throughout the atmosphere and therefore its distribution is controlled by both dynamical and chemical processes. In fact, unlike the WMGHGs, ozone is produced entirely within the atmosphere rather than being emitted into it. Second, it is a very strong absorber of short wavelength UV radiation.

The ozone layer's absorption of this UV radiation leads to the characteristic increase of temperature with altitude in the stratosphere and, in consequence, to a strong resistance to vertical motion. As well as ozone's role in climate it also has more direct links to humans: its absorption of UV radiation protects much of Earth's biota from this potentially damaging short wavelength radiation. In contrast to the benefits of stratospheric ozone, high surface ozone values are detrimental to human health. The distribution of ozone in the atmosphere is maintained by a balance between photochemical production and loss, and by transport between regions of net production and net loss. A number of different chemical regimes can be identified for ozone. In the upper stratosphere, the ozone distribution arises from a balance between production following photolysis of molecular oxygen and destruction via a number of catalytic cycles involving hydrogen, nitrogen and halogen radical species. The halogens arise mainly from anthropogenic ODSs. In the upper stratosphere, the rates of ozone destruction depend on temperature and on the concentrations of the radical species. In the lower stratosphere, reactions on aerosols become important. The distribution of the radicals can be affected by heterogeneous and multiphase chemistry acting on condensed matter. At the low temperature of the wintertime polar lower stratosphere, this is the chemistry that leads to the ozone hole.

The large-scale circulation of the stratosphere, known as the Brewer-Dobson circulation, systematically transports ozone poleward and downward. Because ozone photochemical reactions proceed quickly in the sunlit upper stratosphere, this transport has little effect on the ozone distribution there as ozone removal by transport is quickly replenished by photochemical production. However, this transport leads to significant variations of ozone in the extra-tropical lower stratosphere, where the photochemical relaxation time is very long and ozone can accumulate on seasonal time scales. Due to the seasonality of the Brewer- Dobson circulation, ozone builds up in the extra- tropical lower stratosphere during winter and spring through transport, and then decays photochemically during the summer when transport is weaker. The columnozone distribution is dominated by its distribution in the lower stratosphere and reflects this seasonality. Furthermore, planetary waves are stronger in the Northern Hemisphere than in the Southern Hemisphere, because of the asymmetric distribution of the surface features that, in combination with surface winds, force the waves.

The stratospheric Brewer-Dobson circulation is stronger during the NH winter, and the resulting extra-tropical build-up of ozone during the winter and spring is greater in the NH than in the SH. Variations in the Brewer- Dobson circulation also influence polar temperatures in the lower stratosphere; stronger wave forcing coincides with enhanced circulation and higher polar temperatures. Since temperature affects ozone chemistry, dynamical and chemical effects on column ozone thus tend to act in concert and are coupled. Human activities have led to changes in the atmospheric concentrations of several greenhouse gases, including tropospheric and stratospheric ozone and ODSs and their substitutes. Changes to the concentrations of these gases alter the radiative balance of the Earth's atmosphere by changing the balance between incoming solar radiation and outgoing infrared radiation. Such an alteration in the Earth's radiative balance is called a radiative forcing. This report, past IPCC reports and climate change protocols have universally adopted the concept of radiative forcing as a tool to gauge and contrast surface climate change caused by different mechanisms. Positive radiative forcings are expected to warm the Earth's surface and negative radiative forcings are expected to cool it. Changes in carbon dioxide provide the largest radiative forcing term and are expected to be the largest overall contributor to climate change. In contrast with the positive radiative forcings due to increases in other greenhouse gases, the radiative forcing due to stratospheric ozone depletion is negative. Halocarbons are particularly effective greenhouse gases in part because they absorb the Earth's outgoing infrared radiation in a spectral range where energy is not removed by carbon dioxide or water vapour.

Halocarbon molecules can be many thousands of times more efficient at absorbing the radiant energy emitted from the Earth than a molecule of carbon dioxide, which explains why relatively small amounts of these gases can contribute significantly to radiative forcing of the climate system. Because halocarbons have low concentrations and absorb in the atmospheric window, the magnitude of the direct radiative forcing from a halocarbon is given by the product of its tropospheric mixing ratio and its radiative efficiency. In contrast, for the more abundant greenhouse gases there is a non-linear relationship between the mixing ratio and the radiative forcing. Since 1970 the growth in halocarbon concentrations and the changes in ozone concentrations have been very significant contributors to the total radiative forcing of the Earth's atmosphere.

Because halocarbons have likely caused most of the stratospheric ozone loss, there is the possibility of a partial offset between the positive forcing of halocarbon that are ODSs and the negative forcing from stratospheric ozone loss. The climate impacts of ozone changes are not confined to the surface: stratospheric ozone changes are probably responsible for a significant fraction of the observed cooling in the lower stratosphere over the last two decades and may alter atmospheric dynamics and chemistry. Further, it was predicted that depletion of stratospheric ozone would lead to a global increase in erythemal UV radiation at the surface of about 3 per cent, with much larger increases at high latitudes; these predicted high latitude increases in surface UV dose have indeed been observed.

Chapter Outline

Our aim in this stage is to review the scientific understanding of the interactions between ozone and climate. We interpret the term 'climate' broadly, to include stratospheric temperature and circulation, and refer to tropospheric effects as 'tropospheric climate'. This is a broad topic; our review will take a more restricted view concentrating on the interactions as they relate to stratospheric ozone and the role of the ODSs and their substitutes. For this reason the role of tropospheric ozone in the climate system is mentioned only briefly. Similarly, some broader issues, including possible changes in stratospheric water vapour and climate-dependent changes in biogenic emissions will not be discussed in any detail. The relationship between ozone and the solar cycle is well established, and the solar cycle has been used as an explanatory variable in ozone trend analysis. An update on stratospheric observations of ozone, ozone-related species and temperature. A process level, the various feedbacks connecting stratospheric ozone and the climate system. The understanding of these processes leads to a discussion of the attribution of past changes in ozone, the prediction of future changes, and the connection of both with climate. Finally, reviews the trade-offs between ozone depletion and radiative forcing, focusing on ODSs and their substitutes.

OZONE DEPLETING SUBSTANCES

CFCs are not the only ozone-depleting chemicals (ODCs). A number of other halocarbon species are capable of destroying ozone in the stratosphere. Halocarbons include the chlorofluorocarbons (CFCs), the hydrochlorofluorocarbons (HCFCs), methylhalides, carbon tetrachloride (CCl_4), carbon tetrafluoride (CF_4), and the halons (bromide species). Although the HCFCs do not contribute significantly to the destruction of the ozone layer, they, along with the other halocarbons are all considered to be powerful greenhouse gases and contribute towards global warming.

Hydro Chlorofluoro Carbons

Hydro chlorofluoro carbons or HCFCs contain chlorine but, unlike CFCs, they also contain hydrogen which causes them to break down in the lower atmosphere (troposphere). They are called transition chemicals because they are considered an interim step between strong ozone-depleters and replacement chemicals that are entirely ozone-friendly. Unfortunately, like CFCs, they are strong greenhouse gases and contribute towards global warming.

Carbon Tetrachloride

Carbon tetrachloride (CCl4), despite its toxicity, was first used in the early 1900s as a fire extinguishant, and more recently as an industrial solvent, an agricultural fumigant, and in many other industrial processes including petrochemical refining, and pesticide and pharmaceuticals production. Recently it has also been used in the production of CFC-11 and CFC-12. It has accounted for less than 8 per cent of total ozone depletion. The use of carbon tetrachloride in developed countries has been prohibited since the beginning of 1996 under the Montreal Protocol.

Methyl Chloroform

Methyl chloroform, also known as 1,1,1 trichloroethane is a versatile, all- purpose industrial solvent used primarily to clean metal and electronic parts. It was introduced in the 1950s as a substitute for carbon tetrachloride. Methyl chloroform has accounted for roughly 5 per cent of total ozone depletion. The use of methyl chloroform in developed countries has been prohibited since the beginning of 1996 under the Montreal Protocol.

Halons

Halons, unlike CFCs, contain bromine, which also destroys ozone in the stratosphere. Halons are used primarily as fire suppressants. Halon-1301 has an ozone depleting potential 10 times that of CFC-11. Although the use of halons in developed countries has been phased out since 1996, the atmospheric concentration of these potent, long-lived ozone destroyers is still rising by an estimated 11 to 15 per cent annually. To date halons have accounted for about 5 per cent of global ozone depletion.

Methyl Bromide

Methyl bromide, another bromine-containing halocarbon, has been used as a pesticide since the 1960s. Today, scientists estimate that human sources of methyl bromide are responsible for approximately 5 to 10 per cent of global ozone depletion.

The Ozone Hole: Antarctica

Why is the ozone hole over Antarctica? That is one of the first questions that comes to mind when people think about the ozone hole. Every winter and spring since the late 1970s, an ozone hole has formed in the stratosphere above the Antarctic continent. In recent years this hole has become both larger and deeper, in the sense that more and more ozone is being destroyed. As summer approaches, the hole repairs itself, only to reform during the following spring.

Measuring the Ozone Hole

The most common ozone measurement unit is the Dobson Unit (DU). The Dobson Unit is named after atmospheric ozone pioneer G.M.B. Dobson who carried out the earliest studies on ozone in the atmosphere from the 1920s to the 1970s. A DU measures the total amount of ozone in an overhead column of the atmosphere. Dobson Units are measured by how thick the layer of ozone would be if it were compressed into one layer at 0 degrees Celsius and with a pressure of one atmosphere above it. Every 0.01 millimetre thickness of the layer is equal to one Dobson Unit. The average amount of ozone in the stratosphere across the globe is about 300 DU (or a thickness of only 3mm at 0oC and 1 atmospheric pressure!). Highest levels of ozone are usually found in the mid to high latitudes, in Canada and Siberia (360DU).

Why is the Hole over the Antarctic

Observed ozone over the British Antarctic Survey station at Halley Bay first revealed obvious decreases in the early 1980s compared to data obtained since 1957. The ozone hole is formed each year when there is a sharp decline (currently up to 60 per cent) in the total ozone over most of Antarctica for a period of about two months during southern hemisphere spring (September and October). Man-made emissions of CFCs occur mainly in the northern hemisphere, with about 90 per cent released in Europe, Russia, Japan, and North America. Gases such as CFCs that are insoluble in water and relatively unreactive are mixed throughout the lower atmosphere and rise from the lower atmosphere into the stratosphere; winds then move this air poleward. Normally, chlorine and bromine is inactive, locked up in stable compounds, and does not destroy the ozone.

However, during the Antarctic winter months (June to August) when the region receives no sunlight, the stratosphere becomes cold enough (-80oC) for high level clouds to form, called Polar Stratospheric Clouds (PSCs). These PSCs provide an ideal catalytic surface on which the chlorine can react with the ozone, thus destroying the ozone layer. This reaction requires sunlight, and therefore only begins when the Sun returns to Antarctica in spring (September to October), before the PSCs have had a chance to melt.

The ozone hole disappears again when the Antarctic air warms up enough during late spring and summer During the southern hemisphere winter, Antarctica is isolated from the rest of the world by a natural circulation of wind called the polar vortex. This prevents atmospheric mixing of stratospheric ozone, thus contributing to the depletion of ozone. Although some ozone depletion occurs over the Arctic, meteorological conditions there are very different to Antarctica and so far have prevented the formation of ozone holes as large as in the southern hemisphere.

The Ozone Hole: Arctic

Unlike the Arctic, the Antarctic is isolated from the rest of the world during the winter and spring by a natural circulation of wind called the polar vortex. This prevents atmospheric mixing of stratospheric ozone, thus contributing to the depletion of ozone. However, even small depletions in the Arctic region would give cause for considerable concern due to the higher populations in the higher latitudes of the northern hemisphere.

Do Ozone Holes form over the Arctic

Arctic ozone depletion has not been as marked as over the Antarctic for two reasons:

1. The stratospheric temperatures are seldom below–80°C due to

frequent exchange of air masses with the mid latitudes;
2. The Arctic air vortex usually dissipates in late winter before sunlight returns to initiate the ozone destruction.

The differences between the two regions result in part from the larger land mass in the northern hemisphere, which causes more activity in the atmosphere. Nevertheless, analysis of satellite data reveals that the loss of ozone in the northern hemisphere is now proceeding faster than previously thought. In 1989, NASA's Airborne Arctic Stratospheric Expedition, the first comprehensive research expedition to explore the Arctic region, found that the Arctic stratosphere in winter has almost as much chlorine monoxide as is found in Antarctica, the same destructive chlorine that causes the Antarctic ozone hole.

While no Arctic ozone losses comparable with those in the Antarctic have occurred, localised Arctic ozone losses have been observed in winter concurrent with observation of elevated levels of reactive chlorine, made available through man-made emissions of CFCs. Ozone losses have increased greatly in the 1990s in the Arctic and in late 1997 were the greatest ever observed, just as to measurements by NASA satellites. The rate of loss in mid-latitudes has reached 8 per cent per decade in late winter and spring, and significant loss is now encroaching on the growing season. This compares with the yearly-averaged global mean decrease in the amount of ozone of about 3 per cent in the last decade.

Climate Change and Ozone Depletion in the Arctic

An ozone hole in the Arctic is expected to grow larger over the coming decades as a result of man-made greenhouse gas emissions which may cause climate change, before recovering after 2020. Loss of ozone in the Arctic by

2020 could be about double what would occur without greenhouse gases.

Though greenhouse gases cause atmospheric warming at the Earth's surface, they cool the stratosphere by trapping more heat below, in the troposphere. Since ozone chemistry is very sensitive to temperature, particularly at -80oC when Polar Stratospheric Clouds can form, this stratospheric cooling may result in more ozone depletion in the Arctic.

Temperatures are slightly warmer in the Arctic than the Antarctic during their respective winter and spring seasons, with the result that ozone loss in the northern hemisphere has been lower than that in the southern hemisphere.

But the Arctic stratosphere has gradually cooled over the last decade, resulting in the increased ozone loss. Computer models predict that temperature and wind changes induced by greenhouse gas emissions may allow a stronger and longer-lasting atmospheric vortex to form above the Arctic, as in the Antarctic, causing an increase in ozone depletion. Because of international controls on the emission of ozone-depleting chemicals, those gases are expected to peak about the year 2000.

However, greenhouse gas emissions continue to increase, despite international efforts to control them, and Arctic ozone depletion may continue to worsen until the 2010s, with two-thirds of atmospheric ozone lost in the most severely affected areas.

National Influences on Stratospheric Ozone

Despite all of our harmful chemicals going into the atmosphere, some people still argue that stratospheric ozone depletion is just part of a natural variation. Natural variations which influence the amount of ozone in the upper atmosphere include solar activity, volcanic eruptions and changes in atmospheric circulation - the planetary winds.

The Sun's Influence on Ozone

Stratospheric ozone is primarily created by ultraviolet (UV) radiation coming from the Sun. The Sun's energy release does vary, especially over the 11-year sunspot cycle. During the active phase of the 11-year sunspot cycle, more ozone is produced with the increased UV coming to Earth. This phenomenon can boost the average ozone concentration over the poles by about 4 per cent, but when this is averaged out over the whole earth, the world average ozone increase is about 2 per cent.

Observations since the 1960s have shown that total global ozone levels have decreased by 1-2 per cent from the maximum to the minimum of a typical cycle. However, since downward trends in ozone levels are much larger than 1-2 per cent, particularly at the higher latitudes, the Sun's output cannot be wholly responsible. Unusual solar activity can cause the ozone levels in the upper stratosphere to be substantially depleted, but since most of the ozone is in the middle stratosphere, the effect on the total ozone column is negligible.

Atmospheric Winds and Ozone

A natural cycle in which prevailing tropical winds in the lower stratosphere vary over a time span of about two years can also influence the amount of ozone in the stratosphere.

A change from easterly flow to westerly flow can bring up to a 3 per cent increase in ozone over certain locations, but it is usually cancelled out when the total ozone of the Earth is averaged.

Volcanic Eruptions and Ozone

Volcanic eruptions are one of the few natural things that can have a diminishing effect on the ozone layer. Large eruptions can potentially inject significant quantities of chlorine (via hydrochloric acid - HCl) directly in the stratosphere where the highest concentrations of ozone are found. However, the vast majority of volcanic eruptions are too weak to reach the stratosphere, around 10 km above the surface. Thus, any HCl emitted in the eruption remains in the troposphere where it is quickly dissolved and washed out by rain. In addition, there is no historical record that shows significant increases in chlorine in the stratosphere following even the most major eruptions. It is also possible that ice particles containing sulphuric acid from large volcanic eruptions may contribute to ozone loss. When chlorine compounds resulting from the break- up of man-made CFCs in the stratosphere are present, the sulphate particles serve to convert them into more active forms that may cause more rapid ozone depletion.

In 1991 Mt. Pinatubo in the Philippines erupted tonnes of dust and gas high into the atmosphere which caused global reductions in the ozone layer for 2 to 3 years. Thus, whilst large volcanic eruptions may increase the rate of stratospheric ozone depletion, it is more probable that the presence of chlorine from man-made CFC emissions is the chief cause of ozone loss in the first instance.

Monitoring Ozone Depletion

In 1974, after millions of tons of CFCs had been manufactured and sold, chemists F. Sherwood Rowland and Mario Molina of the University of California began to wonder where all these CFCs ended up. Rowland and Molina theorised that short waves of ultraviolet radiation from the Sun in the stratosphere would break up CFCs, and that the free chlorine atoms would then enter into a chain reaction, destroying ozone. Many people, however, remained unconvinced of the danger until the mid-1980s, when a severe annual depletion of ozone was first monitored by the British Antarctic Survey above Antarctica.

The depletion above the South Pole was so severe that the British geophysicist, Joe Farman, who first measured it assumed his spectrophotometer must be broken and sent the device back to England to be repaired. Once the depletion was verified, it came to be known throughout the world through a series of NASA satellite photos as the Antarctic Ozone Hole.

Evidence for Stratospheric Ozone Depletion

Laboratory studies, backed by satellite and ground-based measurements, show that chlorine reacts very rapidly with ozone. They also show that the chlorine oxide formed in that reaction undergoes further processes that regenerate the original chlorine, allowing the sequence to be repeated very many times (a "chain reaction"). Similar reactions also take place between bromine and ozone. Many other reactions are often also taking place simultaneously in the stratosphere, making the connections among the changes difficult to untangle. Nevertheless, whenever chlorine (or bromine) and ozone are found together in the stratosphere, the ozone-destroying reactions must be taking place. Observations of the Antarctic ozone hole have given a convincing and unmistakable demonstration of these processes.

Monitoring of Ozone Depletion

There has been much monitoring of the condition of the ozone layer in the last decade since the Antarctic ozone hole was first discovered by the British Antarctic Survey. This has utilised satellites and other groundbased resources that are dedicated to observing the destruction of stratospheric ozone. The main satellite that monitors the ozone is the TOMS (Total Ozone Mapping Spectromcter) satellite. The TOMS satellite measures the ozone levels from the backscattered sunlight in the ultraviolet (UV) range. Another satellite is NASA's UARS (Upper Atmosphere Research Satellite) which was launched in September 1991. This satellite is unique because it was configured to not only measure ozone levels, but also levels of ozone-depleting chemicals. GOME, launched in April 1995 on the ERS-2 satellite, marks the beginning of a long-term European ozone monitoring effort. Scientists expect to receive high quality data on the global distribution of ozone and several other climateinfluencing trace gases in the Earth's atmosphere.

The German Neumayer Antarctic Research Station was completed in March of 1992, which is located on the Ekstsoem Ice Shelf. This ground-based station studies geographical, meteorological, and air chemistry conditions. In 1987, Canada became the first country in the world to focus on the Arctic ozone layer, following the discovery of the ozone hole over the Antarctic. A cross-country network of monitoring stations has kept continuous watch on Canada's ozone layer for more than three decades. The existence of these early records, before any major human influence on the upper atmosphere, is vital to understanding the changes that have occurred in the ozone layer. In the UK, stratospheric ozone levels are monitored every winter and spring at Cambourne, in Cornwall and Lerwick, in the Shetland Isles.

❐

5

Carbon Dioxide in Earth's Atmosphere

The concentration of carbon dioxide in Earth's atmosphere is approximately 390 ppm by volume as of 2010[update], and rising by about 1.9 ppm/yr. Carbon dioxide is essential to photosynthesis in plants and other photoautotrophs, and is also a prominent greenhouse gas. Despite its relatively small overall concentration in the atmosphere, CO2 is an important component of Earth's atmosphere because it absorbs and emits infrared radiation at wavelengths of 4.26 μm and 14.99 μm, thereby playing a role in the greenhouse effect.

The present level is higher than at any time during the last 800 thousand years, and likely higher than in the past 20 million years.

CURRENT CONCENTRATION

In 2009, the CO_2 global average concentration in Earth's atmosphere was about 0.0387 per cent by volume, or 387 parts per million by volume. There is an annual fluctuation of about 3–9 ppmv which roughly follows the Northern Hemisphere's growing season. The Northern Hemisphere dominates the annual cycle of CO_2 concentration because it has much greater land area and plant biomass than the Southern Hemisphere. Concentrations peak in May as the Northern Hemisphere spring greenup begins and reach a minimum in October when the quantity of biomass undergoing photosynthesis is greatest.

SOURCES OF CARBON DIOXIDE

Natural sources of atmospheric carbon dioxide include volcanic outgassing, the combustion of organic matter, and the respiration processes of living aerobic organisms; man-made sources of carbon dioxide include the burning of fossil fuels for heating, power generation and transport, as well as some industrial processes such as cement making. It is also produced by various microorganisms from fermentation and cellular respiration. Plants photosynthesis. They produce the energy needed for this reaction through the photolysis of water. The resulting gas, oxygen, is released into the atmosphere by plants, which is subsequently used for respiration by heterotrophic organisms, forming a cycle.

Many sources of CO_2 emissions are natural. For example, the natural decay of organic material in forests and grasslands, such as dead trees, results in the release of about 220 gigatonnes of carbon dioxide every year. In 1997, Indonesian peat fires were estimated to have released between 13 per cent and 40 per cent of the average carbon emissions caused by the burning of fossil fuels around the world in a single year. Although the initial carbon dioxide in the atmosphere of the young Earth was produced by volcanic activity, modern volcanic activity releases only 130 to 230 megatonnes of carbon dioxide each year, which is less than 1 per cent of the amount released by human activities.

These natural sources are nearly balanced by natural sinks, physical and biological processes which remove carbon dioxide from the atmosphere. For example, some carbon dioxide dissolves in sea water, and some is directly removed from the atmosphere by land plants for photosynthesis. There is a large natural flux of CO_2 into and out of the biosphere and oceans. In the pre-industrial era these fluxes were largely in balance. Currently about 57 per cent of human-emitted CO_2 is removed by the biosphere and oceans. The ratio of the increase in atmospheric CO_2 to emitted CO_2 is known as the airborne fraction; this varies for short-term averages but is typically about 45 per cent over longer periods.

Burning fossil fuels such as coal and petroleum is the leading cause of increased anthropogenic CO_2; deforestation is the second major cause. In 2008, 8.67 gigatonnes of carbon were released from fossil fuels worldwide, compared to 6.14 gigatonnes in 1990. In addition, land use change contributed 1.20 gigatonnes in 2008, compared to 1.64 gigatonnes in 1990. This addition, about 3 per cent of annual natural emissions as of 1997[update], is sufficient to exceed the balancing effect of sinks. As a result, carbon dioxide has gradually accumulated in the atmosphere, and as of 2008[update], its concentration is 38 per cent above pre-industrial levels. Various techniques have been proposed for removing excess carbon dioxide from the atmosphere in carbon dioxide sinks.

PAST VARIATION

The most direct method for measuring atmospheric carbon dioxide concentrations for periods before direct sampling is to measure bubbles of air trapped in the Antarctic or Greenland ice caps. The most widely accepted of such studies come from a variety of Antarctic cores and indicate that atmospheric CO_2 levels were about 260–280 ppmv immediately before industrial emissions began and did not vary much from this level during the preceding 10,000 years. In 1832 antarctic ice core levels were 284 ppmv.

One study disputed the claim of stable CO_2 levels during the present interglacial of the last 10 ka. Based on an analysis of fossil leaves, Wagner et al. argued that CO_2 levels during the period 7–10 ka were significantly higher and contained substantial variations that may be correlated to climate variations.

Others have disputed such claims, suggesting they are more likely to reflect calibration problems than actual changes in CO_2. Relevant to this dispute is the observation that Greenland ice cores often report higher and more variable CO_2 values than similar measurements in Antarctica. However, the groups responsible for such measurements believe the variations in Greenland cores result from in situ decomposition of calcium carbonate dust found in the ice. When dust levels in Greenland cores are low, as they nearly always are in Antarctic cores, the researchers report good agreement between Antarctic and Greenland CO_2 measurements.

The longest ice core record comes from East Antarctica, where ice has been sampled to an age of 800 ka. During this time, the atmospheric carbon dioxide concentration has varied by volume between 180–210 ppm during ice ages, increasing to 280–300 ppm during warmer interglacials.

On long timescales, atmospheric CO2 content is determined by the balance among geochemical processes including organic carbon burial in sediments, silicate rock weathering, and volcanism. The net effect of slight imbalances in the carbon cycle over tens to hundreds of millions of years has been to reduce atmospheric CO_2. The rates of these processes are extremely slow; hence they are of limited relevance to the atmospheric CO_2 response to emissions over the next hundred years.

Various proxy measurements have been used to attempt to determine atmospheric carbon dioxide levels millions of years in the past. These include boron and carbon isotope ratios in certain types of marine sediments, and the number of stomata observed on fossil plant leaves. While these measurements give much less precise estimates of carbon dioxide concentration than ice cores, there is evidence for very high CO_2 volume concentrations between 200 and 150 Ma of over 3,000 ppm and between 600 and 400 Ma of over 6,000 ppm. In more recent times, atmospheric CO_2 concentration continued to fall after about 60 Ma. About 34 Ma, the time of the Eocene-Oligocene extinction event and when the Antarctic ice sheet started to take its current form, CO_2 is found to have been about 760 ppm, and there is geochemical evidence that volume concentrations were less than 300 ppm by about 20 Ma. Low CO_2 concentrations may have been the stimulus that favoured the evolution of C_4 plants, which increased greatly in abundance between 7 and 5 Ma.

RELATIONSHIP WITH OCEANIC CONCENTRATION

The Earth's oceans contain a huge amount of carbon dioxide in the form of bicarbonate and carbonate ions — much more than the amount in the atmosphere. The bicarbonate is produced in reactions between rock, water, and carbon dioxide. One example is the dissolution of calcium carbonate:

$$CaCO_3 + CO_2 + H_2O \rightleftharpoons Ca^{2+} + 2HCO_3$$

Reactions like this tend to buffer changes in atmospheric CO_2. Since the right-hand side of the reaction produces an acidic compound, adding CO_2 on the left-hand side decreases the pH of sea water, a process which has been termed ocean acidification. Reactions between carbon dioxide and non- carbonate rocks also add bicarbonate to the seas. This can later undergo the reverse of the reaction to form carbonate rocks, releasing half of the bicarbonate as CO_2. Over hundreds of millions of years this has produced huge quantities of carbonate rocks. Ultimately, most of the CO_2 emitted by human activities will dissolve in the ocean, however the rate at which the ocean will take it up in the future is less certain.

COMPOSITION OF ATMOSPHERE

The most primitive man—the wildest savage—recognizes a wind when out in it, but it took a Greek philosopher to tell us what it is: air in motion. What motion is we know, at least well enough for practical purposes, but what is air? What is that invisible and odourless something we breathe and therefore call atmosphere, the thing that affords us all our weather perceptions and whose states and conditions we have learned to measure and even to foretell, and what was its origin?

ORIGIN OF THE ATMOSPHERE

To start as nearly as possible at the beginning, how and when did the earth ever get an atmosphere—its gaseous envelope? Well, there are two, and so far as we know, only two basic substances, the electron, a certain extremely minute quantity of negative electricity, plus something else, maybe; and the proton, an equally small quantity of positive electricity, plus something else, also maybe. The neutron or chargeless mass is omitted as so little is known about it. Of the origin of these entities no one has the slightest idea. They can exist separately, in which case they are electrically very active; or variously grouped together in equal numbers, with increase of inertia but almost total loss of electric force, at least on things external.

Every such group that is stable, or even measurably durable, is a chemical element. There are no other elements, and this number is limited. Furthermore, though occurring in unequal quantities, all these elements appear to be distributed throughout the universe.

Most of those known on the earth have been found in the sun, for instance, and we believe the others are there, too, even if in such relatively small amounts as to be difficult of detection. Hence, when that other star, according to one cosmic theory, some three billion four hundred million years ago, passed so near our sun as to drag off from it by tidal action the masses that coalesced into the planets, there were present, and came off together, all the possible elements.

That is, the primordial material of the earth, as it was pulled off from, or out of, the sun consisted of all the elements that now make it up, so that at the very beginning of the independent existence of the earth it had, if not an actual atmosphere, at least the makings of one.

It is believed, by those who hold to this theory that as the earth mass drew together in a molten sphere the heavier and more refractory substances formed mainly the inner core, while the lighter and more volatile elements and compounds formed the rocky shell and the gaseous envelope. In this way, they claim, the earth soon had oceans and an atmosphere of some kind. However, even if all the gases, hydrogen, oxygen, nitrogen, and others, that can combine with various elements and compounds and form solids had then so com- bined, leaving neither water nor air, it is certain that before long the earth would have begun accumulating both.

It is doing so now through every volcano, every fumarole, and every bubbling spring, and must always have done so since before even the first crust began to form. But it seems unlikely that our present atmosphere actually did come entirely from the molten interior, whether by way of volcanic activity or otherwise, because volcanoes do not, so far as we know, give off free oxygen—volcanic gases captured before there has been any chance for admixture with the air show no trace of it. Furthermore this element could not exist uncombined in the presence of hydrogen and sulphur at high temperatures, both of which are abundant in volcanic vapors.

But we have oxygen; where did it come from? It is known that highly developed green plants extract, under the stimulus of light, a great deal of oxygen from carbon dioxide, a gas abundantly emitted by volcanos. But most primitive plants do not, they consume it; hence this promising source of free oxygen appears, on close examination, to be quite uncertain. However, some forms of lower life thrive in the absence of free oxygen and yet, in the presence of light, evolve it from certain of its compounds, especially water.

Again, lightning, which must have occurred from the beginning, frees a little oxygen from water; and so also may ultra-violet light. There have been, then, continuously active means of obtaining free oxygen from its compounds since the beginning of the world. Nevertheless, it seems most likely that at the beginning there was more oxygen present than was necessary to use up the free hydrogen and to combine with the available surface materials—enough to do all this and to have a goodly amount left over as free oxygen of the atmosphere.

Presumably, too, there were present other primitive gases, especially nitrogen, argon, carbon dioxide and water vapour. But whatever the primitive state of the atmosphere when the earth first formed, it may be regarded as practically certain that during the whole of the three billion four hundred million years since that time it has been continuously depleted by combination with many things in and of the crust, and also as continuously repleted by their decomposition. It is always changing, but, except in respect to water vapour, the change is so small in comparison with the whole that we are not ordinarily aware of it.

It has been argued that the could not have retained an atmosphere when molten, or even when dull red. But this is true of only the lightest two gases, hydrogen and helium, and of them only if no other gases were present in large amounts. A deep atmosphere of water vapour, for instance, would catch any light gas that might leave the earth beneath with an escaping velocity. It could escape only if the outer portions of the atmosphere also were quite hot.

At any rate, as soon as a crust, however thin, formed over the earth the supply of heat from beneath was so reduced (the crust being a good insulator) that the upper air necessarily became cool enough to retain the lightest gases. Also water must soon have begun gathering on the surface. Even if, up to this stage, all helium and free hydrogen had been driven wholly away from the earth, a condition that seems unlikely, there has been since then, and still is, abundant opportunity to accumulate both of them—hydrogen from volcanos, and helium from radioactive materials everywhere.

Presumably, therefore, the atmosphere is primitive in part—pulled off from the sun with all the other elements—and in part, at least, certainly regenerated inasmuch as every volcano is an active air factory.

Apparent Simplicity of air

Presumably every one usually thinks of air as being just air, a homogeneous and single thing. Many of us always think of it that way, as the ancients did, when we think of it at all. Indeed so far as its behaviour and most of its physical properties are concerned it shows no obvious complexity. It is just air in motion that is responsible for a thousand familiar things from the stir of a leaf to the wreck of a house; and just air that floats the balloon and sustains the aeroplane. In all these matters it usually is quite satisfactory to regard the air as the single substance it ordinarily seems to be.

Evidence of Complexity

This apparent oneness of air does not, however, extend to all physical processes. When we try to liquefy it, for instance, evidences of its complexity soon become amazingly conspicuous. If untreated air is forced through the cooling coils they quickly become choked with ice; and they still clog up when even the driest air is used, if nothing but the water vapour has been removed— this time with solid carbon dioxide. Then, too, the liquid air itself shows abundant evidences that it is a mixture and not a simple substance like water. What then are the known constituents of the atmosphere, and how and when were they discovered?

Water Vapour

The earliest considerations of the composition of the air that have come down to us are those of the Greeks. In their speculative philosophy on the composition of objects, they considered air to be one of the four elements (fire, earth and water being the other three) that, singly or variously combined, make up all substances. From this it might seem that these Greek philosophers regarded the atmosphere as strictly a single thing—the "element" air. Yet it appears that by "air" they meant anything gaseous, and not necessarily the atmosphere. At any rate, Aristotle 250 years B.C., says very distinctly, in his work on meteorology, that cloud and rain are caused by condensation from the atmosphere of water vapour that had gotten there by the evaporation of water at the surface of the earth. He thus makes it very clear that the air consists of at least two things, and that water vapour is one of them.

Water vapour, then, was the first constituent of the atmosphere to be explicitly recognized. Aristotle mentions it, but it is not certain that this discovery was original with him. However, his is the earliest record we have of it, and for that reason, there being no evidence to the contrary, we regard him as one of the discoverers the earliest one of the constituents of the air.

DELAY OF FURTHER DISCOVERIES

For more than 22 centuries, therefore, and perhaps for much longer, it has been known that the air we breathe consists of at least two things, water vapour and whatever is left after the water is removed. And for more than two thousand years after the days of Aristotle this is all that was known about its composition. Indeed it was practically impossible to push our knowledge of the atmosphere any farther without something of the facilities and methods of the modern laboratory, nor before there had been acquired—very slowly and tediously it was—a fair concept of chemical elements and pure substances. Not until the beginning, then, of the 18th century was it reasonably possible for any constituent of the air to be discovered in addition to water vapour, nor indeed was any discovered until long after that.

The chief obstacles that prevented such discovery for more than a hundred years after enough advance for that purpose had been made in laboratory technique, for that had been adequate from the beginning of the 17th century, were:

- The fixed idea that all gases are alike, all just air, and that any differences between various samples are due only to greater or less modifications of one and the same thing.
- The completely misleading and faulty concept that flame or combustion is the escape of something, phlogiston they called it, from within the burning object.
- The failure to recognize that change of weight incident to strong heating, during combustion, or under any other circumstances, was a matter of importance or had any scientific significance whatever.

A century before any constituent of the atmosphere, except of course water vapour, was recognized and collected in an approximately pure state, and while the faulty notions just listed were still prevalent, two people, working entirely independently, came near to finding one or more of its elements.

The first of these was Robert Boyle a wealthy bachelor, chemist and theologian; discoverer of the fact, known as "Boyle's Law," that doubling the pressure on a gas reduces its volume by one-half, of course for the same temperature. It seems very probable that Boyle would have discovered some of the constituents of the air if he had carried to completion certain experiments that he definitely listed. But there is nothing in his voluminous writings to show that he ever got them beyond the paper stage, despite the fact that, so long as his health permitted, he was a persistent worker.

After many and well-devised experiments Mayow concluded that the air consists of at least two portions, one that supports combustion and sustains life, and another part that does neither. The former he called "fire-air," because it keeps a flame going.

He also said that it consists of "nitro-aerial particles," that is, particles in a gaseous form of the kind that makes a mixture of niter and charcoal, or other combustible, burn, when lighted, in the absence of air—even under water.

All this is true enough, but his proof that the air consists in part of a special constituent, different from all the rest, was not complete. He did not collect "fire-air," the gas we now call oxygen, in a practically pure form and show that it is identical with the "fire-air" of the atmosphere. However, he recognized the incompleteness of some of his arguments, and it seems likely that if he had lived a few years longer his proofs would have been perfected, and our knowledge of the composition of the atmosphere set forward almost a hundred years.

But this near attainment to the goal, and also even the direct route to it, appear to have been lost sight of for nearly a century. Indeed "fire-air," that came so near to being the first constituent of the atmosphere to be discovered, except, of course, water-vapour, and whose properties make it the most conspicuous, turned out to be the very last of all the major ones, save only argon.

Carbon Dioxide

The first of the permanent gases of the air of nearly constant quantity to be clearly discovered was not, as would seem most likely, either oxygen or nitrogen that together make up nearly the whole, but carbon dioxide, that is present as scarcely more than a trace—three parts in ten thousand. At that time alkalies were given to persons suffering with urinal calculi, and certain physicians recommended lime water for the same purpose. Which the view of finding something still better Black undertook investigations with magnesia alba, a form of carbonate, as we now know, or, more exactly, basic carbonate, of magnesium.

On strongly heating this substance a gas is given off, and Black turned his attention particularly to that gas, or air, as all gases were then called. He tried calcining, or burning to a powder, various substances, such as limestone, that we now know to be carbonates, and studying the gas thus obtained. In the end he found that this gas is much heavier than ordinary air, that it will not support life or combustion, that it will recombine with the calx, or powder, produced by the strong heating of the original substance, and that heat will again expel it as before. This gas seems to be fixed, or somehow fastened, in the objects from which it may be obtained.

Nitrogen

The composition of the air in a closed vessel after it no longer would support combustion or life, and to determine the cause of its unwholesomeness. After burning charcoal, phosphorus, or other combustible, in a closed volume of air, as long as possible, he removed the "fixed air" (carbon dioxide), if any had been formed, by means of lime, or an alkali, all in accordance with the previous investigations of Black, and then examined the remaining gas. He showed that this residue is not ordinary air because it supports neither life nor combustion, and that it is not fixed air for the alkalies do not absorb it. He called this residue "mephitic air," because it does not support life.

We now know and say that this residue had been obtained by burning out the oxygen of the confined air and then absorbing the carbon dioxide thus produced. And we know, too, that this residue, this "mephitic air," was nearly pure nitrogen. But at that time oxygen had not yet been discovered, and of course Rutherford could not talk in terms of things and chemical reactions then unknown. He did know, however, that it was obtained by combustion in an inclosed or limited volume of ordinary air, and therefore concluded, after the philosophy of his day, that it was atmospheric air combined with, or modified by the addition of, phlogiston—a mysterious fire substance whose escape from an object commonly is manifested by flame. Nevertheless, and no matter what his ideas as to its nature, Rutherford did obtain reasonably pure nitrogen and did record some of its properties, and therefore may be regarded as the discoverer of this constituent of the air, the most abundant of all.

Oxygen

Almost immediately after the discovery of nitrogen, the other major constituent of the atmosphere, oxygen, was independently found by Joseph Priestly. Priestley was a preacher by occupation and a chemist for recreation. At one time, while waiting for the building of the parsonage to be finished, he had the good fortune to live next door to a brewery—good fortune, because he was induced thereby to take up the chemistry of gases, his chief avocation for many years, on which he published several volumes, and which made his name famous. In the course of this work he obtained a red powder by heating mercury in the presence of air, and then on more strongly heating this powder with a burning lens he got a gas which supported combustion much better than ordinary air. He had gotten oxygen, as we now know, by first burning mercury to an oxide and then decomposing that oxide by raising it to a high temperature.

This element of the atmosphere he called dephlogisticated air. Combustible things burned in it readily; their phlogiston, or fire principle, rushed into it as air into a vacuum. It was air, that is, a gas, but air deprived of phlogiston. But regardless of what he called it Priestley had discovered a new constituent of the atmosphere, the one we now call oxygen, the one that sustains life and supports combustion. On the other hand, Scheele was a professional chemist, so completely absorbed in his subject that he had little time for anything else, whether occupation or diversion.

His incentive to study the air was his wish to solve the riddle of fire, a thing essential to so many chemical processes. He found substances, such as phosphorus, which on burning reduced the volume of the confined air (at the same temperature and pres sure) in which they were burned by about one- fifth. In all cases the remaining air would not support combustion.

He also found many ways of getting a gas that would support combustion far better than ordinary air, that was heavier than the air left after combustion, and that made burnt air indistinguishable from ordinary air when mixed with it in the proportion of one volume to four, or thereabouts. The residual or burnt air, the nitrogen, essentially, as we now name it, he called "vitiated air." The other portion, the part that supports combustion, he called "fire-air," just as Mayow had called it a century before.

Scheele, like Priestley and every other chemist of his day, except the immortal and tragic Lavoisier, interpreted fire phenomena in terms of phlogiston. He therefore considered "fire-air," which we call oxygen, to be a combination of phlogiston and a subtle acid substance. But whatever his opinions may have been as to its possible composition, he did find this constituent of the atmosphere, and that too even before it was found by Priestley.

Another student of the atmosphere that must be mentioned in connection with the discoveries of its composition was Henry Cavendish. He may not generally be credited with the discovery of any one of these constituents, and yet he appears to have found nitrogen about the same time that Rutherford did, if not earlier, by the simple process of passing the same confined air back and forth over red hot charcoal and then removing the fixed air (carbon dioxide) with an alkali.

He did not publish this—he appeared always to be indifferent about publishing his investigations—but described it in a letter to Priestley in 1772, the year Rutherford published his discovery of mephitic air.

Lord Rayleigh had been making careful measurements of the densities of different gases, and thus came upon the fact that the nitrogen of the atmosphere, or rather that residual then regarded as pure nitrogen, was a little denser than the nitrogen obtained from any one of several chemical compounds.

This was capable of three or four different interpretations, one of which, then regarded as the least likely, was the presence in the atmosphere of an unknown gas denser than nitrogen. At that stage Sir William Ramsay joined Lord Rayleigh in a common attack on this problem. Each tried removing from the air the then known constituents, using entirely different methods for getting rid of the nitrogen, and each found conclusive evidence of the presence in the atmosphere of an unknown gas to the extent of about one per cent. It proved to be chemically inert, and for that reason was called argon.

Other Inert Gases

During the years 1895-1898, that is, immediately after the discovery of argon, four additional inert gases, helium, neon, krypton and xenon, were found by Sir William Ramsay and his assistant. Travers, to be constituents of the atmosphere. Helium was found as that portion of the atmosphere that is still gaseous while the rest has been chilled to liquids or solids; and the others, neon, krypton and xenon by fractional distillations of large quantities of liquid air and liquid argon, impure, of course.

Hydrogen

Lord Rayleigh and others also found free hydrogen always present in the air, but in amounts that varied from, roughly, one part in 1,000,000 to one 5,000. Perhaps this gas can not be regarded as a permanent component of the atmosphere in the same sense that oxygen and nitrogen are, but rather as an accidental and variable impurity. At any rate it is irregularly added to the air, at least by volcanoes, and more or less irregularly removed from it.

Traces and Impurities

Varying traces of ammonia, nitric and nitrous acids and their compounds, sulphuric and sulphurous acids and their compounds, oxides of nitrogen, hydrogen dioxide, and ozone are among the innumerable things always in the air if not of it. Minute particles of sea salt from evaporating spray, fine earth dust caught up by winds, pollen of every description and spores of many kinds are among the coarser and ever present pollutions of the atmosphere from the earth beneath. Shooting stars furnish a continuous, though invisible, shower of dust the world over from outer space.

Also radioactive products of radium, thorium and other elements are pouring into the atmosphere continuously and contributing to the maintenance of its electrical state. This state means, in part, around 20,000 electrified particles, or ions, per cubic inch in the lower air. The number of ions 60 miles or so above the surface is far greater, at least 1,600,000 per cubic inch, as we infer from the phenomena of radio communication.

In short, while the things of the atmosphere are but few the things in it are innumerable. All these latter are relatively very small in amo t, but some of them are exceedingly important. The salt particles and some others are essential to condensation and rainfall; the ammonia and other nitrogen compounds, brought down by precipitation, add much to the fertility of the soil; the ions of the high atmosphere make distant radio communication possible; and ozone, also in the upper air, shields us from that portion of the ultra violet radiation that would destroy our eyesight, as we are now constituted, and otherwise do us irreparable harm.

PERCENTAGE COMPOSITION OF PURE DRY AIR

If the atmosphere is freed from all its numerous impurities and also freed from water vapour, there will remain pure dry air whose percentage composition at the surface of the earth is, very closely:

Nitrogen	78.03
Oxygen	20.99
Argon	0.9323
Carbon dioxide	0.03
Hydrogen	0.01
Neon	0.0018
Helium	0.0005
Krypton	0.0001
Xenon	0.000009

The amount of water vapour in the atmosphere, or better, perhaps, the amount of the water vapour constituent of the atmosphere, varies from scarcely more than a trace, at extremely low temperatures, to at least 5 per cent by volume on the hottest and most humid days.

At and above the height of 5 miles, say, the amount of water vapour always is small, even when saturation obtains, owning to the very low temperatures at these levels. That is, water vapour is confined almost wholly to the lower atmosphere. Its average value, the world over, is such that if it were all condensed it would be the equivalent of a layer of water about one inch deep over the entire earth.

Carbon Monoxide

Carbon monoxide (CO) is a colourless, odourless, poisonous gas produced when fuels containing carbon are burned where there is too little oxygen. It also forms as a result of burning fuels at too high a temperature. It burns in air or oxygen with a blue flame and is slightly lighter than air.

In the presence of an adequate supply of O_2 most carbon monoxide produced during combustion is immediately oxidised to carbon dioxide (CO2). However, this is not the case in spark ignition engines in motor cars, especially under idling and deceleration conditions.

Thus, the major source of atmospheric carbon monoxide is road transport. Smaller contributions come from processes involving the combustion of organic matter, for example in power stations and waste incineration.

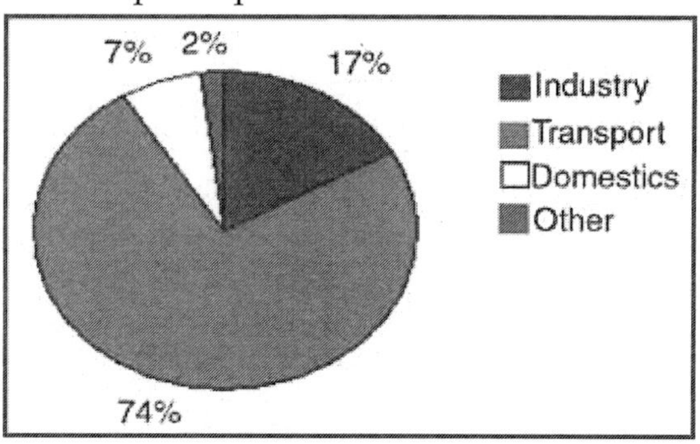

Fig. *Carbon Monoxide Emissions in The UK 4758 Thousand Tonnes*

Natural background levels of carbon monoxide fall in the range of 10-200 parts per billion. Levels in urban areas are highly variable, depending upon weather conditions and traffic density. Concentrations are generally less than 10 parts per million (ppm) but can be as high as 500 ppm.

Carbon monoxide is poisonous when inhaled because it combines with haemoglobin, the oxygen-carrying substance in red blood cells. The hemoglobin then cannot take up oxygen from the air. Lack of oxygen causes cells and tissues to die.

Cars

The motor vehicle engine emits many types of pollutants including nitrogen oxides (NOx), volatile organic compounds (VOCs), carbon monoxide (CO), carbon dioxide (CO_2), particulates, sulphur dioxide (SO_2) and lead. Emissions of such pollutants may lead to poor air quality, posing environmental and health risks. Individually, a vehicle engine is not a particularly important source of pollution. Collectively however, they represent a major source of air pollutants inmost developed countries, including the UK.

Emissions are related to use of the engine, mainly the fuel type and the temperature of fuel combustion. If the engine is efficient, then the products of combustion will be mainly carbon dioxide and water. However, at low speeds and idling engines are inefficient and therefore the products of incomplete combustion dominate, for example carbon monoxide and VOCs from petrol engines, and carbon monoxide, VOCs, particulates and smoke from diesel engines. As the temperature of combustion increases, the efficiency of conversion to carbon dioxide and water increases. However, impurities in the fuel such as nitrogen are oxidised to nitrogen dioxide. At high temperatures atmospheric nitrogen (N_2) is also oxidised to nitrogen dioxide.

Nitrogen oxides pollution from vehicles can be significantly reduced by fitting a catalytic converter to the exhaust system. This is a relatively low cost method of pollution control which has little effect on vehicle performance and fuel consumption. All new cars sold in Britain from January 1993 onwards have catalytic converters.

Diesel fuel contains more energy per litre than petrol and coupled with the fact that diesel engines are more efficient than petrol engines, diesel cars are more efficient to run. Diesel fuel contains no lead and emissions of the regulated pollutants (carbon monoxide, hydrocarbons and nitrogen oxides) are lower than those from petrol cars without a catalyst. However, when compared to petrol cars with a catalyst, diesels have higher emissions of nitrogen oxides and much higher emissions of particulate matter.

Emissions from cars are greatest when an engine is cold. On a cold day a petrol car may take up to 10km to warm up and operate at maximum efficiency; a diesel car may only take 5km. Consequently, diesel cars produce less unburned fuel during a cold start, which will result in lower emissions of carbon monoxide and hydrocarbons. Diesel cars could make a significant impact on air quality in urban areas where most cold starts occur, especially when it is considered that a catalyst on a petrol car would take several minutes to reach its operating temperature.

Overall, diesel cars emit less hydrocarbons, carbon monoxide and lead pollution than petrol cars, but produce more noxious gases and significantly more particulates.

Catalytic Converters

Since January 1993, all new petrol-driven cars sold in the European Union (EU) have been fitted with a catalytic converter (CAT). This is made up of a very thin layer of platinum group metals on a honeycomb structure. The surface area of a typical 3-way CAT covers the equivalent of two football pitches. As exhaust gasses pass through the catalyst a chemical reaction occurs which converts carbon monoxide (CO), volatile organic compounds (VOCs, including hydrocarbons) and nitrogen oxides (NOx) to less harmful compounds (water, nitrogen and carbon dioxide).

To work most effectively, a catalytic converter needs to reach an optimum temperature. It may not reach this in a short journey. Devises to pre-warm the catalyst are being developed which improve the overall performance of catalytic converters.

The use of catalytic converters leads to a dramatic reduction in the emissions of carbon monoxide, hydrocarbons and nitrogen oxides. However, they also result in an increase in carbon dioxide (CO_2) emissions, which do not cause a problem for urban air quality, but may contribute to global warming. The efficiency of a CAT can be as high as 90 per cent.

Oxidation catalysts may be fitted to either petrol or diesel cars. The catalyst oxidises the pollutants formed by incomplete combustion to carbon dioxide and water, and is effective for hydrocarbons and carbon monoxide. However, they do not reduce nitrogen oxides emissions.

Clean Air Acts

In response to the Great London Smog of December 1952, the Government introduced its first Clean Air Act in 1956. This Act aimed to control domestic sources of smoke pollution by introducing smokeless zones. In these areas, smokeless fuels had to be burnt. The Clean Air Act focused on reducing smoke pollution, but the introduction of cleaner coals and the increased usage of electricity and gas actually helped to reduce sulphur dioxide levels at the same time. In addition, power stations were relocated to more rural areas. As a consequence, air pollution in cities was dramatically reduced.

The Clean Air Act of 1968 introduced the basic principle for the use of tall chimneys for industries burning coal, liquid or gaseous fuels. At the time of this legislation it was recognised that smoke pollution could be controlled, but that sulphur dioxide removal was generally impracticable. Hence, the higher the chimney, the better the dispersal of the air pollution.

Cleaner Fuels

To replace pollutant fuels (petrol and diesel), cleaner alternative fuels are currently being developed. They include: compressed natural gas (CNG); liquefied petroleum gas (LPG); city diesel; hydrogen; alcohol fuels; and battery operated vehicles.

On a cycle representing congested urban traffic, both LPG and CNG have lower emissions of carbon monoxide (CO) than for petrol-powered vehicles. Indeed, emissions of carbon monoxide from CNG-powered vehicles are of the same order as those emitted by diesel vehicles. However, emissions of hydrocarbons from CNG vehicles are relatively high because of methane, the major component of natural gas.

Although methane is a small contributor to the formation of low level ozone it is a major factor in global warming. Emissions of nitrogen oxides and particulates from both LPG- and CNG-powered vehicles are significantly lower than those from diesel vehicles. Moreover, emissions of nitrogen oxides from CNG vehicles are half those from equivalent petrol-engined vehicles.

City diesel is a petroleum-based lower emission diesel developed in Sweden but now available in many European Countries including the UK. Exhaust emissions from vehicles fuelled with city diesel compare favourably with exhaust emissions from equivalent vehicles fuelled with conventional diesel. The main benefit of city diesel is that its combustion reduces particulate emissions by 34 - 84 per cent depending on engine type and type of particulate measured. An additional benefit of city diesel is that it is a low sulphur fuel, which is necessary for the optimum running of oxidation catalytic converters.

Whilst pollution from road transport may be reduced by using alternative fuels to petrol and diesel, ultimately it is cheaper to improve conventional fuels than to use many of the alternatives, and no investment is needed for new storage tanks and service stations.

Dispersion

Many atmospheric factors influence the way air pollution is dispersed, including wind direction and wind speed, type of terrain and heating effects. To better understand how atmosphere processes can affect ground level pollution, atmospheric conditions can be described simply as either stable or unstable, where the stability is determined by wind (which stirs the air) and heating effects (which cause convection currents). Atmospheric stability affects pollution released from ground level and elevated sources differently.

In unstable conditions, ground level pollution is readily dispersed thereby reducing ground level concentrations. Elevated emissions, however, such as those released from a chimney, are returned more readily to ground level, leading to higher ground level concentrations.

Stable conditions mean less atmospheric mixing and therefore higher concentrations around ground level sources, but better dispersal rates, and therefore lower ground level concentrations, for elevated plumes.

CARBON MONOXIDE IN NATURE

Carbon monoxide, with the chemical formula CO, is a colourless, odourless and tasteless, yet highly toxic gas. Its molecules consist of one carbon atom and one oxygen atom, connected by a covalent double bond and a dative covalent bond. It is the simplest oxocarbon, and can be viewed as the anhydride of formic acid (CH_2O_2).

Carbon monoxide is produced from the partial oxidation of carbon-containing compounds; it forms in preference to the more usual carbon dioxide (CO_2) when there is a reduced availability of oxygen, such as when operating a stove or an internal combustion engine in an enclosed space.

Carbon monoxide has significant fuel value, burning in air with a characteristic blue flame, producing carbon dioxide. Despite its serious toxicity, it was once widely used (as the main component of coal gas) for domestic lighting, cooking and heating, and in the production of nickel. Carbon monoxide still plays a major role in modern technology, in industrial processes such as iron smelting and as a precursor to myriad products.

Molecular Properties

The carbon monoxide molecule consists of one atom of carbon and one atom of oxygen, covalently bonded by a double bond and a dative covalent bond. Its bond length is 0.1128 nm. The effects of atomic formal charge and electronegativity result in a small bond dipole moment with its negative end on the carbon atom.

The reason for this, despite oxygen's greater electronegativity, is that the highest occupied molecular orbital has an energy much closer to that of carbon's p orbitals, meaning that greater electron density is found near the carbon.

In addition, carbon's lower electronegativity creates a much more diffuse electron cloud, enhancing the polarizability. This is also the reason that almost all chemistry involving carbon monoxide occurs through the carbon atom, and not the oxygen.

The bond length of CO is consistent with a partial triple bond, and the molecule can be represented by three resonance structures:

FORMS OF OXYGEN AND THEIR DISTRIBUTION

Oxygen also is peculiar in its distribution, and it occurs in three different forms. All but about one part in 400,000 is ordinary oxygen, or, in the language of the chemist, diatomic oxygen. Triatomic oxygen, or ozone, occurs chiefly beyond the highest clouds, its greatest density being, apparently, at the levels of 20 to 30 miles. It occurs in the lower air only as a trace. In the highest air, 60 miles and more above the surface, the oxygen appears to be at least partly monatomic, according to the spectrum of the aurora.

Importance of Ozone

Every one knows that life would be impossible if there were no ordinary or diatomic oxygen in the atmosphere, and that without it nearly if not quite all vegetable life also would be impossible. But it is not so well known, though equally true, that animals, including the human species, could not exist as now constituted if the air did not contain a small amount of triatomic oxygen or ozone.

And yet, paradoxical as it may if seem, ozone were just a little more effective in its goodness, again life, as now constituted, could not last. These surprising facts come about in this way: The radiation from the sun includes not only every colour, that is, the whole of the visible spectrum, but also extends indefinitely beyond the red into the long-wave-length invisible region, and likewise, in the opposite direction, well beyond the limit of the violet.

But in this ultra violet portion of the spectrum the radiation from the sun ceases far short of the limit to which that from an electric arc, for instance, can be followed; and ceases, not because no radiation beyond that limit is given out by the sun, but because it is absorbed by the ozone in the upper atmosphere. Now, much of that particular radiation which ozone absorbs is destructive to the eye, and when intense probably injurious to other tissues as well.

On the other hand, it also absorbs a large part of the radiation that is effective in preventing rickets, but, and this is of the utmost importance, it does not absorb quite all of this antirachitic portion. Enough is let through to keep us in proper health. This particular limiting, then, of the solar radiation that reaches the earth is amazingly located. If it was a little farther out in the ultra violet, eyes, as now constituted, could not have developed; and if a little nearer the visible, again, animals, as they now are, could not have come into being. Of course ozone, in moderate amount, presumably was in the air long before there were animals of any kind to be affected by its action on solar radiation. This radiation, therefore, was not adapted to them, but they developed in adaptation to it; nevertheless the fit is so close and of such a surprising nature as to give us decided pause for thought.

Origin of Ozone

Ozone is produced by the action of extreme ultra violet light on oxygen, and therefore at great heights; a little where the oxygen is very rare, then more and more with decrease of height and increase of the oxygen supply until by absorption the ozonizing rays are considerably enfeebled. As the density of the oxygen increases the intensity of the effective radiation decreases, hence the rate of production of ozone by this process must be very slow in the outermost air, increase to a maximum with decrease of height and then rapidly fall off at still lower and lower levels.

The total range of ozone, top to bottom, probably is not less than 100 miles, with its maximum concentration around 25 miles, perhaps, above the surface, and yet all told, and despite its great importance, it is the equivalent of a layer of this gas only about one-tenth of an inch thick at atmospheric pressure and room temperature. Ozone is produced also by electric discharges through oxygen. Lightning certainly produces some ozone, and it may be that the auroras form it too, though the great heights at which they occur, 60 miles and more above the earth, seem to render this conclusion doubtful.

Distribution of Ozone

As already stated, we know that at most there is only a trace of ozone in the atmosphere up to the level of the highest clouds, and that it exists to an appreciable extent at considerably greater heights. But this is not all. The total amount of ozone vertically over one to the limit of the air appears to increase, in general, with increase of latitude; to be greater during winter and spring than summer and fall; and to be greater in winds from high latitudes than in those of the opposite direction. This complicates the problem of the origin of ozone. If it is produced by ultra violet radiation alone why should it not be most abundant in tropical regions, and elsewhere in the summer time?

If we surmise that it is produced largely by auroral discharges, how, we ask, can these discharges, 60 miles up and more, reach a sufficient supply of oxygen? And if there is enough oxygen at these great heights how can the ozone subsequently become concentrated at the intermediate levels? These are some of the questions to be answered by future observations and studies.

Mass of the Atmosphere and of its Constituents

From the known percentages of the several constituents of dry air, given above, their molecular weights, and various other pertinent facts such as the amount of water vapour present, height of the barometer, volume of land above sea level, and distribution of temperature with height, it is easy to compute the approximate mass of the atmosphere as a whole and of each of its several gases. The results are given in the following table, in which the factor 108 means: Add eight ciphers, or multiply by 100,000,000.

Since the values in this table were determined more or less independently it could not be expected that the percentages found of the constituents would add up to exactly 100, nor that the sum of the computed masses of the several parts would precisely equal the mass of the whole. These deviations, however, are very small—probably within the present limits of experimental errors.

Table: *Mass of the Atmosphere and of its Constituents in Tons (2000 Pounds)*

Substance	Volume per cent dry air, at surface	Total mass
Total atmosphere		56,328,000 × 108 tons
Dry air	100.00	56,181,850 " " "
Nitrogen	78.03	42,684,725 " " "
Oxygen	20.99	12,782,647 " " "
Argon	0.9323	682,125 " " "
Water vapor		146,150 " " "
Carbon dioxide	0.03	23,874 " " "
Hydrogen	0.01	1,423 " " "
Neon	0.0018	759 " " "
Krypton	0.0001	141 " " "
Helium	0.0005	88 " " "
Ozone	0.00006	33 " " "
Xenon	0.000009	19 " " "

The numbers here given that express the masses of the atmosphere and its several constituents are useful as quantitative values and for exact comparisons, but so great, even though in terms of tons, that we can form no distinct conceptions of them—they are just awfully big! A clearer idea may be gotten from the fact that the total mass of the atmosphere is the equivalent, roughly, of that of a block of granite a thousand miles long, a thousand miles broad and half mile thick; while the least abundant of the constituent, xenon, if loaded on cars, 19 tons to the car, would freight a train reaching 40 times around the earth along a great circle, and which, traveling 20 miles all hour, would be 6 years in passing any point on the road.

EARTH'S ATMOSPHERE

The future of any science can always be painted, what with the exponential rate of most human progress, as a brilliant picture in the most glowing colours. Of course this is particularly true of the great science of aerology, for its spacious labouratory, measurably colossal and immeasurably complicated, is the whole broad air blanket that is the Earth's atmosphere.

Any worker in the field of aerology comes across many factors of aerological equipment, technique, and procedure that are quite apparently in need of considerable improvement. To each intelligent and thoughtful worker there occur many possible solutions — 'inventions,' if you like — only a few of which he will ever have time to try out in actuality. What are the things that need improving in present-day aerology? What are the inventions, of equipment or method, that loom on the horizon of this, Earth's highest frontier?

Better Instruments and Observations

The most common aerological instruments are those devoted to measuring atmospheric pressure, temperature, and humidity. As to barometres, little if any future improvement over the present delicate and efficient instruments can be expected, and various difficulties combine to make any further advances along this line of doubtful value. We already have micro-barometres and micro- barographs which are more sensitive than we can practically use. An ultra-sensitive micro-barograph shows slight pressure variations occasioned by opening and shutting doors and windows, changes in wind directions or speed, earth tremors, and various other common natural occurrences which have nothing to do with the present or future state of the weather. Hence a less sensitive instrument is, for weather purposes, often more useful.

In the thermometre field, however, quite a few improvements can easily be envisioned. Perhaps the most obvious one is the more extensive use of remote-indicating thermometres of the electric type, in which the temperature- sensitive element can be located up to several hundred feet away from the indicating dial or recording pen. The widespread use of these electric gadgets, now entirely too expensive but potentially much cheaper if produced in mass, will permit better exposure of the sensitive element.

At present the thermometre shelters at many weather stations are located in untypical places (from a temperature viewpoint) such as valley bottoms, city districts, and so on. The result is that many local temperatures, as measured today, are inaccurate by several degrees. The same sort of improved technique might be applied to the measurement of atmospheric humidity — a characteristic fully as important as temperature in weather work. Here the key gadget might be a remote-indicating hygrometre of the electric-resistance type. Another technical development of possible usefulness might be a 'searchlight thermocouple' similar to those used by astronomers for determining the temperature of planets, stars, and so on. For aerological use these thermocouples might be pointed at clouds, portions of the upper air, and so on.

As indicated earlier in this book, the measurement of wind direction and speed at various places leaves much to be desired. The important wind for weather analysis, in most regions, is the 'gradient' wind or 'geostrophic' wind found at levels between five hundred and fifteen hundred feet above the ground or water surface. Tall radio towers and other high, open structures near weather stations should be equipped with remote-indicating anemometres wherever possible. At larger radio stations, it might be worth building three hundred-foot steel-lattice towers just for temperature, humidity, and wind measurements above the ground-effect level. A promising gradient- wind-measuring gadget is a small mortar of about 1½-inch bore (similar to the Lyle gun used in alongshore lifesaving), which fires a small paper projectile straight upward. The projectile is filled with black powder and fused so as to explode at several hundred feet altitude. The resulting cloud of white smoke can be observed with optical instruments and its direction and speed of drift of course give the desired data on the gradient wind.

Another promising gadget is the cloud-thickness metre. In simplest form, this is merely a sensitive photometre (a good photographic exposure metre will do) pointed straight upward. With the sun at equal elevation, the observed light-intensity reading is of course roughly proportional (inversely) to the total thickness of clouds above the station — a bit of information often useful in weather work, particularly in aviation. Still another gadget is the visibility metre. Where visibility is the same thing as air transparency, a measurement can be made by some such method as having two white circles or other targets of unequal size, placed at such distances that they appear equal in size (or angular diametre) to the observer. The comparative clearness with which each device can be seen is then a measure of atmospheric transparency or visibility. Another recent development in the study of the upper air employs polarized light. An aerologist at Pennsylvania State College has found that, by measuring sky polarization with a special instrument, it is possible to sense the increase in size of condensation nuclei which usually precedes any sudden formation of lower clouds or fog.

For measuring ceiling (or height of cloud base) at night, vertical searchlights are in common present use; and whenever the ceiling is below a mile, and provided the searchlight is powerful and accurately focused, this method leaves little to be desired. In the daytime ceilings are now measured by means of small balloons that are supposed to rise at a fixed rate. But this method has many disadvantages, and something better is badly needed. By using high-power radiation in some suitable portion of the visible or invisible spectrum, or by using polarized light, it might be possible to devise a sort of 'day ceiling light.' An- other possibility in day-ceiling indicators, which was suggested to me in 1937 by Mr. George A. Wolcott of the Weather Bureau, might be a rocket that rises vertically and leaves a visible trail of coloured smoke. The point where this smoke trail intersects the cloud base, sighted on by an inclinometre at the other end of a base line, would of course give accurately the existing ceiling, provided the rocket actually rises vertically. Or perhaps a smoke-trail projectile could be fired vertically out of a smoothbore mortar.

Radio can be used in several fields of weather investigation. One that any amateur can try out is based on the fact that a lightning flash produces radio waves (mostly low-frequency around 200 kilocycles) that can be heard in any suitable receiver. Hence an increase of this 'crash static' usually means thunderstorm activity in the vicinity; and by using directional receivers along fixed base lines, thunderstorms can be located and followed in their wanderings. A somewhat similar radio-direction-finding technique can be used to locate and follow small tropical hurricanes (which usually have a good deal of thunderstorm activity near their centres) from fixed goniometric stations along the coast.

Short-wave radio can be used to some extent in locating and studying cloud masses in the upper air, the boundaries between air masses of different temperature or humidity, and other phenomena of importance in weather work. For example, in 1932. an amateur operating an ultra-short-wave (five- metre) portable transmitter on the summit of Pike's Peak found that clouds drifting between him and the other station on the plateau far below would cut out the radio signal entirely, apparently blocking it off by reason of refraction, reflection, absorption, or some other effect. Work of this sort has also been carried on by two ultra-short-wave radio stations operating under the auspices of Harvard University on Mount Washington in New Hampshire and Whiteface Mountain in upstate New York. Since 1935 Mr. A. W. Friend of Harvard University has carried out extensive radio explorations of the troposphere with radio frequencies around 2400 kilocycles (about 125 metres). The apparatus can locate at will various discontinuity surfaces in the troposphere: temperature inversions (including the stratosphere base); the sloping planes of warm, cold, and upper fronts; tops and bottoms of cloud layers; turbulence layers. From a radio-echo sounding of existing turbulence levels aloft, Mr. Friend can even tell a departing aviator at what altitudes he is likely to find excessively rough and bumpy air.

Another radio development which fills a real need in weather work is the fixed robot radio weather station or 'ray-mete.' This consists of various meteorological instruments which measure and record temperature, humidity, wind, and so on. By means of a complicated gadget the observed weather data is transmitted by radio to a remote recording station which may be several hundred miles distant. The robot station, somewhat in the manner of an automatic lighthouse, can be left 'untouched by human hands' for six months at a time. Obviously the development of this technique towards full practical usefulness will permit the establishment of robot weather stations at many important but inaccessible locations — high mountainpeaks, selected points in arctic wastes and tropical deserts or jungles; ocean positions one or two hundred miles off our Atlantic and Pacific coasts.

Completer Analysis and Forecasting

One thing that impresses any scientific worker in these days — or, in fact, any intelligent and widely read layman — is the increasing mechanization and mathematication of all human science. Much of this is doubtless necessary in our present world, but some of it is perhaps not so necessary and a good deal of it is rather mournful to contemplate, particularly if you are the type of aerologist or layman who likes to look at clear skies and clouds and sunrises and moonsets and rainbows and halos for their own sake, as things of surpassing ethereal beauty, without any thought of the complex mathematical formulae which may enter into the precise laws governing them. Nevertheless mathematics has a very definite place in aerology or any other branch of physics, and any serious worker in these fields must travel the long and tedious road that runs from algebra through calculus.

One might, for example, write pages of near-poetic description in order to give a clear idea of what a thunderstorm is like. But a scientist might describe the same thunderstorm somewhat as follows:

$$Cp \log T - AR \log p + 0.6_{23} \frac{re_m}{p^r} = K$$

$$AR \log \frac{P^o}{p^r} + \left[\frac{1+K}{p^r} - \frac{r}{p^0} \right] \frac{0.6_{23^e{}_m}}{T} - \frac{K_q}{T} = 0$$

$$Cp \log T - AR \log p + \frac{0.6_{23}(r+K)e_m}{p^r} = K$$

And his succinct statement might be more useful, to all the bright young men studying at Massachusetts Tech. and California Tech. in preparation for work with the Weather Bureau, the Army and Navy, commercial airlines, and all the other weatherwatchful agencies in the modern American scene.

We have not often in this book bedeviled the lay reader with technical equations such as those above, nor will we in the future. In my own opinion there has been a tendency in recent years for mathematics to run away with physical science -something that often looks strangely like a small tail wagging a large dog. Mathematics is at best a shadowland of the mind which does not, and never will, represent concrete reality in any adequate and satisfying fashion. And while we are involved in this digression, it might be further said that the importance of physical science in general and certain aspects of technology in particular have perhaps been somewhat exaggerated. Far more important than all physical science, at least in the future, is social science — not how to invent more machines and methods, civil and military, but how to use intelligently and constructively, for general human benefit, those machines and methods that we already have in abundance.

A powerful new technique in weather analysis called 'isentropic analysis' was developed in 1937 at Massachusetts Tech. by Professor Carl-Gustave Rossby and his associates. The general idea of this method, in the briefest terms, is as follows: Potential temperature usually increases with altitude. Hence it should be possible by means of many atmospheric soundings over a given region to construct a diagram showing the conformations of the various isentropic (equal-potential-temperature) surfaces above that region. Further, choosing one isentropic surface in middle altitudes it should be possible to draw a sort of contour map of it, in the same way that one draws a topographic contour map of some smaller region of the earth's surface.

In the troposphere, the potential temperature normally increases upward at a rate of about five Centigrade degrees per kilometre. It increases southward at about the same rate as the ordinary temperature. Thus the troposphere may be subdivided in a great number of thin isentropic strata (layers of constant potential temperature) which gradually descend towards the equator.

Isentropic-analysis charts are usually drawn and studied by aerologists in connection with the atmospheric cross-sections described elsewhere in this book. The practical end result of these isentropic charts is, in particular, a knowledge of where in the upper air great tongues of dry air are invading territory formerly occupied by moist air, and where similar tongues of moist air are penetrating into regions formerly dry.

This latter sort of débâcle among 'prohibition forces' aloft may be a very significant piece of writing on the weather-portent wall — as appeared in the case of the tragic New England hurricane. As Major Floyd B. Wood of the United States Army Aerological Service points out, 'The importance of the isentropic conception is that wind currents of unsaturated air move in planes parallel to the isentropic surfaces, rather than in horizontal planes (the common conception). Hence the importance of knowing the isentropic contour, and wind direction and velocity relative thereto.' One weather-forecasting possibility which has for years been a perennial dream among aerologists is the matter of foretelling weather many days or weeks or months or even years in advance. Quite a few amateurs have dabbled in this field with more or less popular success — success achieved mostly because human beings are rather gullible, by and large, with a tendency to remember spectacular successes and forget less colourful fail- ures. Some of the really scientific work in this field has been concerned with advances and recessions in the polar ice caps. Most of all, this seasonal weather variability probably depends on changes in the character and intensity of the all-powerful radiation from the sun — changes evidenced by sun spots, bright flares or faculae on the sun; by magnetic storms, auroras, and the like on earth. This fascinating field in solar-terrestrial research.

A more immediate development in long-range forecasting is the endeavor to forecast weather for several days — or say up to one week — in advance. The Germans have done considerable work in this field, to which their plodding, methodical minds are ideally suited. Perhaps the most noteworthy work has been accomplished in the United States, under the auspices of the Weather Bureau — by Professor Hurd C. Willett, of the Massachusetts Institute of Technology and his associates. In Willett's words:

It has been realised since the daily analysis of the Northern Hemisphere map was started at M.I.T. in the fall of 1936, that variations of the general circulation pattern, as defined by the intensity and position of the principal centres of action of the northern hemisphere, largely determine the persistent anomalies of the meteorological elements which are observed on this hemisphere.

In less technical language, the usual North American 'centres of action' directly control the advances and retreats and conflicts of air masses and fronts that cause our day-to-day weather. An unusual intensification of the Canadian- Alaskan high in winter, for example, usually causes a series of cold waves some days or weeks later over much of the United States. An extraordinary intensification of the Bermuda-Azores high in summer usually wafts a series of hot waves over the same region.

HIGH-PILOTS AND RAY-PILOTS

As early as 1935, it was apparent to some aerologists that long-range aviation was gradually moving upstairs. Stratosphere flying, or at least substrarosphere flying, was talked of as something bound to come within the next few years. The standard pilot balloon in ordinary use at that time, leaving the ground with a diametre of something like two feet and a free lift of about five ounces, usually disappeared from the observer's view before it had ascended more than four or five miles, even in the clearest weather; and at night, the small electric light or candle lantern which it carried aloft seldom permitted observations above altitudes of two or three miles. These altitudes, indeed, were likely to be considerably less if visibility were low or if upper winds were unusually high. Yet people were talking of flying the long-range, high-speed air transports of the future at altitudes of five or six miles; and frequent wind reports along the course ahead had become an essential part of every airline pilot's bag of navigating tricks.

On the Stratosphere Expedition of May-July, 1935, in South Dakota, described at some length in the opening chapter, wc carried out some preliminary experiments in the direction of bigger and better pilot balloons and more powerful light sources for night use with these balloons. For the sake of simplicity, both these balloons and their night lights are here called 'highpilots.' In that clear North Dakota air, using two theodolite stations located about five miles apart at an altitude of about a mile above sea level, we got some remarkable observations extending to unusually great heights.

Continuing this 'high-pilot' research project back in the East during the next two years, some of us in the Army Aerological Service made hundreds of other balloon runs to high altitudes at Bolling Field in the District of Columbia and at Aberdeen Proving Ground in Maryland. At about this time, in 1936, there became available an improved large pilot balloon (inflatable at the ground to a diametre of nearly four feet and a free lift of more than two and a half pounds), manufactured by an entirely new process — a cheaper and lighter balloon than any that had ever before been available for this sort of work.

The immediate effect of the new lightness in the balloon itself was to make possible a rise rate considerably faster than had been possible with the smaller standard pilot balloons — or even with the large, heavy sounding balloons that we had used in North Dakota. (This faster rise rate of course extended the altitude of possible upper-wind observation; because the balloon had less time in which to blow away to great distances.) We were soon able, in fact, by inflating the new balloon to a free lift of about forty ounces, to achieve rise rates around 400 yd/min - just twice as fast as the rise rate of the smaller pilot balloon in general use at that time.

Whenever the daytime skies were clear and visibility was good we were able to follow these balloons to their ultimate bursting altitude somewhere between forty and fifty thousand feet — unless exceptionally high upper winds blew the balloon away to leeward unusually fast. Most of the time the balloon would reach bursting altitude at a distance of twenty or thirty miles from the main theodolite station. One afternoon we put a balloon run to thirty-six thousand feet on the airways teletype — to the consternation of the Washington operator, in whose estimation, apparently, such things just were not done, or at least should not be done in the existing state of the art. By hanging a five- or six-ounce weight on a ten-yard string beneath the balloon, it was possible (paradoxically enough) to speed up the rise rate still further — apparently because the balloon was pulled out slightly into a pear-like, or roughly streamlined, shape.

In scores of night balloon runs, trying for record-breaking altitudes under clear skies (clear for the East, that is, but nothing like as clear as Western skies in comparable weather), we tested various sorts of light sources — little acetylene burners supplied with gas from a smaller balloon hanging below the main lifting balloon; red pyrotechnic flares (similar to railroad danger signals) that were specially made up for us by a fireworks factory so as to burn with constantly increasing candlepower; three-watt electric lights consisting of a 1½-candlepower automobile bulb lit by six medium-sized flashlight cells; and a large tissue-paper lantern, cylindrical in shape and about one foot in diametre by three feet high. (Inside the lantern, attached to the bottom ring, was a compact group of four standard candles, all burning merrily away with a combined radiance of five or six candlepower. Both the acetylene light and the pyrotechnic flare never continued burning above twenty thousand feet — perhaps owing to oxygen shortage, or extreme low temperature.

The large tissue-paper candle lantern was, of course, a clumsy thing to carry around; and particularly difficult to launch, when the surface winds were high and gusty, without setting the whole lantern afire and destroying it then and there. (The lantern was completely closed with tissue paper at both top and bottom and, strangely enough, continued to burn brightly during the ascent with this small amount of ventilation.) Nevertheless this old-fashioned candle lantern, once successfully launched, proved to be our best light source for dependably reaching high altitudes at night. Our highest Eastern night observation with the four-candle lantern extended to 25,500 feet altitude above station, checked by double theodolite on the twentieth minute when the lantern was about thirteen miles distant and encountering seventy-mile winds. Later we developed a candle unit in which four candles started the run, and were arranged by connecting wicks to light four more shorter candles after ten or fifteen minutes, giving twice the candlepower when the light was far distant and faint to the vision, but there was no opportunity to test this improved lantern in actual ascents before the whole project ended.

In many states having large forest areas it is illegal to send up candle lanterns or other burning devices with balloons. For this reason, and also to get a light source that could be launched without difficulty in high and gusty surface winds, we carried out many test night ascents with the three-watt electric light. With seven or eight volts applied to the six-volt automobile bulb, it naturally burned blue-white with excessive brightness at the start; but as the balloon ascended and the minutes went on this brightness steadily diminished, and this diminution of the light itself, together with the increasing distance out as the balloon sailed away to leeward, made the traveling light speck among the stars harder and harder to follow as it gained increasing altitude. However, the electric-light source performed quite well on the whole. The best night altitude reached with it was 23,250 feet, when the light was about ten miles away in seventy-mile winds eighteen minutes after leaving the ground. A more powerful electric-light source, employing eight flashlight cells to give an effective starting voltage of ten volts or more on the six-volt bulb, and consequently a blinding light of 'photo-flood' intensity, was designed and tested on the ground.

Unfortunately, there was not time to make actual balloon ascents with this more powerful light before the entire research project had to be finished. When visibility is poor, or when fog or clouds obscure the sky, it is all too evident that no day balloons (however large) and no night balloon lights (however powerful) will serve to get the upper-wind reports that are often so important to aviators and aerologists. In dense haze infra-red radiation, rather than visible light, might help — but even infra-red is powerless to penetrate heavy clouds. For such work a still lower-frequency (or longer wave length) form of electro-magnetic radiation must be used — radio waves. And even the shorter radio waves may suffer absorption, reflection, or other distortion from clouds or fronts.

Hence the all-weather upper-wind-finder of the future, now scarcely beyond the labouratory stage, is the radio pilot balloon or 'ray-pilot.' While some earlier work had been done with impractical spark transmitters, the first important ray-pilot experiments with vacuum-tube transmitters were made about 1923 by Colonel W. R. Blair of the United States Army Signal Corps. By about 1928 the Signal Corps Labouratories had further developed the ray- pilot into more effective and reliable form. The transmitter, weighing scarcely more than a pound, was surprisingly powerful and efficient. Many tests were carried out on wave lengths near 125 metres (2400 kilocycles), including careful radio-tracking of the balloon by means of specially developed direction-finding receivers on the ground, the radio bearings being checked by theodolite observations. It was found that the balloon could be easily followed, with useful azimuth readings, out to distances of ten miles or so from the ground stations. Within five miles' distance, the radio bearings were accurate to less than half a degree. No radio readings of the balloon's angular elevation were attempted with this equipment — its assumed altitude was figured from the average rise rate.

With the development of ultra-short-wave radio in the nine teen-thirties, it was apparent that this new frequency region offered many advantages for ray-pilots. The transmitters, as well as their antennae, could be smaller and lighter.

The ultrashort-wave region was free from the high-power interference that cluttered up the longer-wave channels. Unlike the longer waves, ultra-short waves do not bend around the bulge of the earth or around hills or mountains, but follow a path, even as light does, that is straight or nearly straight. But this straightline propagation was no disadvantage in ray-pilot work — for the rapidly rising balloon would always be above any terrestrial hills or other obstacles.

Present in our 1935 Stratosphere Expedition were many distinguished scientists, and among the radio experimenters was Mr. Harry Diamond of the National Bureau of Standards. He soon joined forces with our pilot-balloon crew, and the result was a pioneering experiment in the field of ray-pilots and raysondes. We used no less than three large sounding balloons, all tied together in a loose cluster. Several feet beneath the balloons we hung a small radio transmitter operating at 108 megacycles (2.8 metres) with a radiated power of about one twentieth watt, and weighing about forty ounces complete. This weight was of course excessive, but the transmitter was rather hastily thrown together by Mr. Diamond and his assistant Mr. Lester from various available parts and batteries that happened to be in the Stratocamp.

This large balloon cluster was released from one of our baseend theodolite stations (though not without some difficulties occasioned by gusty surface winds) under a clear blue South Dakota sky at about noon. The little radio transmitter bobbed around erratically on the end of the string, and its signal consequently swung and faded badly. Also, the balloons went away to the northward in the lower levels, whereas the receiving beam antenna was pointed southeast. Nevertheless, Mr. Diamond followed and measured the minute signal from the balloons at this ultra-high, and heretofore difficult, frequency for thirtyeight minutes on his superheterodyne receiver.

At this time the balloon cluster was at an altitude of about forty-one thousand feet above sea level, and distant about twelve miles to the eastnortheast. After the radio signal died away the visual balloon run continued, followed by both our theodolites for over three hours. Between the eighty-seventh and eighty-eighth minutes one balloon burst, and the remaining two, weighted by the radio transmitter and the empty balloon, began to settle slowly downward. The balloons had faded out to the naked eye after about half an hour; then, after three quarters of an hour, they had reappeared to the naked eye coming westward again in a marked east-west drift around 40,000 feet. At the eighty-seventh minute, the altitude of the balloons was about 60,100 feet above sea level, and the distance out about fifteen miles to the east-northeast.

At the one hundred and ninety-sixth minute, when the balloons were finally lost, they had descended to an altitude of about 15,000 feet, and were about twenty-six miles out to the east. The balloons and the radio transmitter were picked up about 3.30 P.M. of the same day by a farmer living about forty miles to the east.

After returning East to the Bureau of Standards, Diamond continued his pioneering work with ultra-high-frequency ray-pilots (as well as with the ray- sondes to be described in the next section). Using the remarkably high frequency of 200 megacycles (1½ metres), he carried out a variety of outdoor labouratorytype tests which clearly indicated the directions that future ray- sonde progress must take. The azimuth of a small ultrahigh-frequency radio transmitter drifting with the rising balloon can be found by two ground receiving stations at the ends of a suitable baseline. Diamond determined, with an accuracy of something between one half and one degree. This accuracy is sufficient in the practical work of determining upper winds in or above the clouds.

More important, the Bureau of Standards experiments showed that, by means of suitable antenna arrangements, the elevation angle of the ray-pilot could be determined by radio with an accuracy of one degree or less. Assuming that the ray-pilot was also acting as a ray-sonde — sending out a continuous record of the pressure it encountered, and hence its altitude — its distance away from the ground station could be readily computed. Thus the plotting of the invisible balloon's position in space could all be done from one ground receiving station, with even better accuracy than that possible where two azimuth-bearing stations were used.

Many technical difficulties are involved in this ultra-shortwave ray-pilot research, and the practical perfection of an efficient and reliable ray-pilot is still in the future. In most of Mr. Diamond's experiments so far, the transmitter has been raised by means of an elevator at a short distance from the receiver, rather than by a balloon rising freely in the atmosphere, in order that complete control over each step in the experiment could be maintained. Yet the technical difficulties are being gradually overcome, and before many years it should be possible for aerologists to know the upper winds by day or by night, come clear or come storm.

RISING RADIO ROBOTS

Visits to the stratosphere are to become a daily occurrence. Lieutenant W. H. Wenstrom of the Signal Corps, United States Army, has been making a study of radio sounding balloons at the California Institute of Technology and has indicated the feasibility of meteorological observations by the use of such instruments. At present the cost of production of a radio sounding balloon exceeds the cost of an airplane flight, but Lieutenant Wenstrom is very optimistic about the future use of radio sounding balloons. During bad weather when airplane flights cannot be undertaken, meteorological observations are most desirable. A radio balloon can be released in any weather and it will rise to much greater heights than could be negotiated by an airplane, with an instantaneous reception of the data on the ground....

Only four incomplete series of experiments with ray-sondes that actually ascended had been carried out (by Blair in the United States, Bureau in France, Duckert in Germany, and Moltchanoff in Russia); and although there was a great hullabaloo about large stratosphere balloons, the whole ray-sonde idea was terra incognita to most American scientists.

As far back as 1875 by a Dutch instrument-maker named Olland. In the briefest possible terms, the Olland telemeteorograph was as follows: Imagine a watch face, around which a sweep-second hand makes a complete revolution in some definite interval, such as every thirty seconds.

Between twelve o'clock and four o'clock on the watch face is another hand, also pivoted at the centre; the position of this hand (over its own third of the dial) is regulated by the atmospheric pressure. Between four o'clock and eight o'clock is a third hand, similarly regulated by the atmospheric temperature. Between eight o'clock and twelve o'clock there is a fourth hand indicating the atmospheric humidity.

The thirty-second sweep-hand, driven by clockwork at uniform speed around the watch face, makes electric contact with the other hands in the following order:

- A fixed reference at twelve o'clock;
- The pressure hand;
- The temperature hand;
- The humidity hand.

Each electric contact would transmit a 'tick' signal by wire (or equally well by radio, which was still to come in Olland's day). The time intervals between the successive ticks, noted by ear or recorded by a suitable machine, would give a continuous record of the pressure, temperature, and humidity being experienced by the aerograph, whether it was installed on the ground (at a remote robot weather-reporting station), or rising through the troposphere with a sounding balloon. In 1935, partly at the suggestion of some of us in the Army Aerological Service, Blue Hill Observatory of Harvard University started work on ray-sondes using the

Olland telaerographic principle — the simplest, and in some ways the best, principle that will ever be applied to this work. Doctor K. O. Lange developed the aerograph part of the Blue Hill ray-sonde; Mr. A. E. Bent did the radio work. By the summer of 1936 this Harvard ray- sonde was emerging from the labouratory stage and had proved itself in scores of actual upper-air soundings. One of these, was reported to the Chief Signal Officer of the Army as follows:

In the Blue Hill instrument, the flat disk is replaced by a small insulating drum not much larger than a thimble. The pressure, etc. indicating arms move up and down on this drum like the pens of a standard recording instrument. The rapidly moving radial contact arm is replaced by a conducting silver thread wound in helical form on the small drum.

Three quadrants of the cylinder are used respectively for the pressure, temperature, and humidity pens. The reference pen occupies the fourth quadrant, near the end of the drum; the helical silver thread makes one reference contact, and a small silver rivet makes another. This general principle and ingenious construction are believed to be the best so far developed anywhere.

The entire radio-sonde is enclosed in a cellophane casing, in order to absorb and retain solar heat at high altitudes. The radio part of the Blue Hill radio- sonde is also distinctive, in that ultrahigh frequencies are used. The very small, light transmitter operates on 68 megacycles (about 4.5 metres wave length) with an output of about one fourth watt.

As tested by the Weather Bureau in 1937-38, the Blue Hill raysonde weighed less than one pound complete, including the aerograph proper, the miniature radio transmitter, and its power supply.

Also embodied in the improved Harvard instrument was a cheap, reliable, and reasonably cold-proof clock mechanism (weighing only two ounces) which drove the sweep-second hand (or rather, the small revolving drum) one revolution in thirty seconds. In order to adapt the instrument to night soundings, the 'green house' cellophane casing had been replaced by a heat- insulating balsa-wood box.

Another early form of ray-sonde in wide use by 1936 was the instrument devised by Professor Vaisala in Finland. Unlike the clock-driven aerograph element of the Harvard ray-sonde, Vaisala's radio-aerograph used as driving power a sort of small windmill, or arrangement of anemometre cups, which would steadily revolve in the downdraft caused by the steadily rising balloon. Another interesting feature of the Vaisala ray-sonde was a lead-acid-type storage battery of extremely light weight (three ounces) that would give constant voltage and could be recharged for repeated use. In the autumn hurricane season of 1936, an expedition from the Massachusetts institute of Technology used twenty Vaisala ray-sondes and ten Harvard raysondes for hurricane research in the West Indies.

Nineteen hundred and thirty-six was, in fact, a banner year in American ray-sonde development. In response to a request from the United States Navy Department, Mr. Diamond and some of his associates in the Bureau of Standards were bringing to bear on the problem some heavy scientific artillery even as the first Harvard ray-sondes were rising into the New England skies. The specifications set up by the Navy Aerological Service were as follows:

The instrument was to have a unit cost of about twenty-five dollars in reasonable quantities, weigh about one pound, and provide pressure indications in the range of from 1000 to 200 millibars with an accuracy of indication of one millibar, temperature indications in the range of from +30 to -70° C. accurate to within one degree, and humidity indications in the range of from zero to 100 per cent relative humidity accurate to within 3 per cent. Consideration was to be given in the design to the provision of emitted signals suitable for use by radio direction finders on the ground in order to permit determination of upper-air wind conditions.

Mr. Diamond and Mr. Hinman and Mr. Dunmore did not like the idea of either a clockwork or an air fan to drive their aerograph mechanism. Hence they reverted to an earlier system which had been used by Duckert some years before — in which the pressure-indicating element itself served as the driving mechanism. As the balloon rises, the pressure of course decreases, steadily and more or less uniformly; thus the moving arm of the pressure element can be made to serve as an electric switch to close successive contacts, emitting characteristic radio signals. In the words of Diamond, Hinman, and Dunmore:

The pressure arm, moving continuously in one direction as the balloon ascends, switches the frequency of the audio oscillator to correspond alternately to the values of the temperature and of the humidity encountered. The alternate change-overs from one set of frequencies to the other indicate that the pressure arm is just reaching or is just leaving one of the intermediate contacts and his attained definite deflection positions. When the pressure arm reaches successive fifth conducting segments, the frequency of the audio oscillator attains predetermined fixed values which positively identify these contacts so that they may serve as index marks for the absolute-pressure scale....

At the ground station was a complicated radio receiver-recorder which would automatically plot a complete adiabatic diagram as the balloon ascended — temperature and humidity plotted as abscissas against pressure as ordinates.

In complete form the Diamond-Hinman ray-sonde, as manufactured by Friez of Baltimore and adopted for standard radiosounding use throughout the United States and its territories and its surrounding oceans by the United States Weather Bureau and the Navy Department in 1939-40, weighed about two pounds. The radio frequency employed as a carrier wave was about 65 megacycles (about 4.6 metres); the various modulation frequencies which between 10 and 190 cycles, so that the radio signal from the balloon sounded somewhat like an organist practicing various low notes. Below the balloon itself, on a ten-yard length of heavy twine, hung a small silk parachute, a springy shock-absorber ring, and the radio-aerograph itself, with the short transmitting antenna hanging downward from the aluminum-foil-covered, corrugated-pasteboard box. Usually the balloon would burst at about fifteen miles' altitude, the parachute would open, and the aerograph would descend safely to earth. On the average, about half of the ray-sondes released in settled regions are picked up and returned (in response to a reward of several dollars); and about two thirds of these, or one third of those released, can be repaired and used again. At sea or in remote regions, a released ray-sonde is of course lost for good; but the important upper-air data are nevertheless secured.

By 1940 'RAOBS' (ray-sonde observations) were being made nightly from something like forty stations scattered all over the United States, not to mention about ten ships of the Navy in various adjacent waters. In the Southwest, for example, radio soundings were being made at San Diego and Oakland in California, Phoenix in Arizona, and Ely in Nevada. Nevertheless the older method of airplane observations or 'APOBS' had not been entirely discontinued. Useful in many military and naval situations, the APOBS were still being made at some Army and Navy stations.

Still another American ray-sonde has been developed by Doctor L. F. Curtiss of the Bureau of Standards, using ultrashort-wave radio together with a ray-mete based on the original Olland telemeteorographic principle. Doctor Curtiss has also pioneered, along with such renowned American scientists as Karl Compton and Andrew Millikan, radio-sounding-balloon ascents far up into the stratosphere with the object of measuring cosmic radiation at these extreme altitudes. One of the Curtiss ray-sondes, ascending from Washington, D.C., was followed by a Navy airplane from the Anacostia Station. The balloon of course rose far above the airplane's utmost ceiling — to a height of about seventeen miles; but as the burst bubble plummeted downward, the watchful Navy airman got close to it about two miles up and by diving steeply followed it down to the ground, where it was recovered, about thirty miles from Washington, by a surprised Maryland farmer. One of the Curtiss ray-sondes reached an altitude of about twenty-five miles or 132,000 feet above sea level—probably a world's record for any sort of balloon.

Still other Bureau of Standards scientists — Coblentz and Stair — have used ray-sondes for measuring the intensity of ultra-violet radiation from the sun in the stratosphere. Their instruments employed a photoelectric ultra- violet-metre connected so as to operate a miniature radio transmitter of the Diamond-Hinman type. Several ascents were made, which showed high temperatures and lethally intense ultra-violet radiation in the upper stratosphere. The importance of this research in such fields as long-range weather forecasting can scarcely be exaggerated, for the solar radiation which varies most widely and which probably affects earthly weather most violently is precisely in this ultra-violet region of the spectrum.

Another possible radio-robot device for sounding the blue depths above might be a sort of oversized model airplane — say with a wingspread of something like ten or fifteen feet and, horsepower of fifty or so — designed for rapid climb, radiocontrolled, and equipped with a ray-mete that would send back the essential p.,t., and f. data as the pilotless plane sounded upward. Many little planes of this general type are now in use as anti-aircraft artillery targets in progressive armies.

A further development in radio-controlled and radio-reporting gadgets for sounding the troposphere might use the helicopter principle. Professor von Karman of California Tech. (one of the world's authorities in aeronautics) suggested to me in 1934 that the idea should be feasible. The actual sounding device might be an oversized model helicopter, equipped with two large, oppositely revolving lift propellers on the same shaft, and balloonlike shock-absorber feet to cushion the landing jolt on descent. With efficient helicopter design, one horsepower will lift a total weight of something like fourteen kilograms (about thirty pounds). Hence for moderate-altitude sounding in all sorts of weather, one might envision a little whirligig six or eight feet in diametre, run by a miniature gasoline motor of about one horsepower and weighing complete (structure, motor, gas supply, radio control, and aerological ray-mete) about thirty pounds. The little spinner could take off vertically from the roof of the weather station, climb at 600 to 1000 ft/min straight upward under a ground aerologist's control, and perhaps return shortly, still under control, to a bouncy landing on the airport.

❐

6

Climate Change

Climate change is a long-term change in the statistical distribution of weather patterns over periods of time that range from decades to millions of years. It may be a change in the average weather conditions or a change in the distribution of weather events with respect to an average, for example, greater or fewer extreme weather events. Climate change may be limited to a specific region, or may occur across the whole Earth. In recent usage, especially in the context of environmental policy, climate change usually refers to changes in modern climate. It may be qualified as anthropogenic climate change, more generally known as global warming or anthropogenic global warming (AGW).

TERMINOLOGY

The most general definition of climate change is a change in the statistical properties of the climate system when considered over periods of decades or longer, regardless of cause. Fluctuations on periods shorter than a few decades, such as El Nino, do not represent climate change. The term sometimes is used to refer specifically to climate change caused by human activity; for example, the United Nations Framework Convention on Climate Change defines climate change as "a change of climate which is attributed directly or indirectly to human activity that alters the composition of the global atmosphere and which is in addition to natural climate variability observed over comparable time periods." In the latter sense climate change is synonymous with global warming.

Causes

Factors that can shape climate are climate forcings. These include such processes as variations in solar radiation, deviations in the Earth's orbit, mountain-building and continental drift, and changes in greenhouse gas concentrations. There are a variety of climate change feedbacks that can either amplify or diminish the initial forcing. Some parts of the climate system, such as the oceans and ice caps, respond slowly in reaction to climate forcing because of their large mass. Therefore, the climate system can take centuries or longer to fully respond to new external forcings.

Plate Tectonics

Over the course of millions of years, the motion of tectonic plates reconfigures global land and ocean areas and generates topography. This can affect both global and local patterns of climate and atmosphere-ocean circulation. The position of the continents determines the geometry of the oceans and therefore influences patterns of ocean circulation. The locations of the seas are important in controlling the transfer of heat and moisture across the globe, and therefore, in determining global climate. A recent example of tectonic control on ocean circulation is the formation of the Isthmus of Panama about 5 million years ago, which shut off direct mixing between the Atlantic and Pacific Oceans.

This strongly affected the ocean dynamics of what is now the Gulf Stream and may have led to Northern Hemisphere ice cover. During the Carboniferous period, about 300 to 360 million years ago, plate tectonics may have triggered large-scale storage of carbon and increased glaciation. Geologic evidence points to a "megamonsoonal" circulation pattern during the time of the supercontinent Pangaea, and climate modeling suggests that the existence of the supercontinent was conducive to the establishment of monsoons. The size of continents is also important. Because of the stabilizing effect of the oceans on temperature, yearly temperature variations are generally lower in coastal areas than they are inland. A larger supercontinent will therefore have more area in which climate is strongly seasonal than will several smaller continents or islands.

Solar Output

The sun is the predominant source for energy input to the Earth. Both long- and short-term variations in solar intensity are known to affect global climate. Three to four billion years ago the sun emitted only 70 per cent as much power as it does today. If the atmospheric composition had been the same as today, liquid water should not have existed on Earth. However, there is evidence for the presence of water on the early Earth, in the Hadean and Archean eons, leading to what is known as the faint young sun paradox. Hypothesized solutions to this paradox include a vastly different atmosphere, with much higher concentrations of greenhouse gases than currently exist Over the following approximately 4 billion years, the energy output of the sun increased and atmospheric composition changed, with the oxygenation of the atmosphere around 2.4 billion years ago being the most notable alteration. These changes in luminosity, and the sun's ultimate death as it becomes a red giant and then a white dwarf, will have large effects on climate, with the red giant phase possibly ending life on Earth. Solar output also varies on shorter time scales, including the 11-year solar cycle and longer-term modulations. Solar intensity variations are considered to have been influential in triggering the Little Ice Age, and some of the warming observed from 1900 to 1950. The cyclical nature of the sun's energy output is not yet fully understood; it differs from the very slow change that is happening within the sun as it ages and evolves. While most research indicates solar variability has induced a small cooling effect from 1750 to the present, a few studies point towards solar radiation increases from cyclical sunspot activity affecting global warming.

Orbital Variations

Slight variations in Earth's orbit lead to changes in the seasonal distribution of sunlight reaching the Earth's surface and how it is distributed across the globe. There is very little change to the area-averaged annually averaged sunshine; but there can be strong changes in the geographical and seasonal distribution.

The three types of orbital variations are variations in Earth's eccentricity, changes in the tilt angle of Earth's axis of rotation, and precession of Earth's axis. Combined together, these produce Milankovitch cycles which have a large impact on climate and are notable for their correlation to glacial and interglacial periods, their correlation with the advance and retreat of the Sahara, and for their appearance in the stratigraphic record.

Volcanism

Volcanism is a process of conveying material from the crust and mantle of the Earth to its surface. Volcanic eruptions, geysers, and hot springs, are examples of volcanic processes which release gases and/or particulates into the atmosphere. Eruptions large enough to affect climate occur on average several times per century, and cause cooling (by partially blocking the transmission of solar radiation to the Earth's surface) for a period of a few years. The eruption of Mount Pinatubo in 1991, the second largest terrestrial eruption of the 20th century (after the 1912 eruption of Novarupta) affected the climate substantially. Global temperatures decreased by about 0.5 °C (0.9°F). The eruption of Mount Tambora in 1815 caused the Year Without a Summer. Much larger eruptions, known as large igneous provinces, occur only a few times every hundred million years, but may cause global warming and mass extinctions. Volcanoes are also part of the extended carbon cycle. Over very long (geological) time periods, they release carbon dioxide from the Earth's crust and mantle, counteracting the uptake by sedimentary rocks and other geological carbon dioxide sinks. The US Geological Survey, however, estimates are that human activities generate more than 130 times the amount of carbon dioxide emitted by volcanoes.

Ocean Variability

The ocean is a fundamental part of the climate system. Short-term fluctuations (years to a few decades) such as the El Nino–Southern Oscillation, the Pacific decadal oscillation, the North Atlantic oscillation, and the Arctic oscillation, represent climate variability rather than climate change. On longer time scales, alterations to ocean processes such as thermohaline circulation play a key role in redistributing heat by carrying out a very slow and extremely deep movement of water, and the long-term redistribution of heat in the world's oceans.

Human Influences

In the context of climate variation, anthropogenic factors are human activities that change the environment. In some cases the chain of causality of human influence on the climate is direct and unambiguous (for example, the effects of irrigation on local humidity), while in other instances it is less clear. Various hypotheses for human-induced climate change have been argued for many years. Presently the scientific consensus on climate change is that human activity is very likely the cause for the rapid increase in global average temperatures over the past several decades. Consequently, the debate has largely shifted onto ways to reduce further human impact and to find ways to adapt to change that has already occurred.

Of most concern in these anthropogenic factors is the increase in CO_2 levels due to emissions from fossil fuel combustion, followed by aerosols (particulate matter in the atmosphere) and cement manufacture. Other factors, including land use, ozone depletion, animal agriculture and deforestation, are also of concern in the roles they play – both separately and in conjunction with other factors – in affecting climate, microclimate, and measures of climate variables.

ASTRONOMICAL THEORY AND ABRUPT CLIMATE CHANGES

Throughout the 19th and 20th centuries, a wide range of geomorphology and palaeontology studies has provided new insight into the Earth's past climates, covering periods of hundreds of millions of years. The Palaeozoic Era, beginning 600 Ma, displayed evidence of both warmer and colder climatic conditions than the present; the Tertiary Period was generally warmer; and the Quaternary Period showed oscillations between glacial and interglacial conditions. Louis Agassiz developed the hypothesis that Europe had experienced past glacial ages, and there has since been a growing awareness that long-term climate observations can advance the understanding of the physical mechanisms affecting climate change. The scientific study of one such mechanism–modifications in the geographical and temporal patterns of solar energy reaching the Earth's surface due to changes in the Earth's orbital parametres has a long history. The pioneering contributions of Milankovitch to this astronomical theory of climate change are widely known, and the historical review of Imbrie and Imbrie calls attention to much earlier contributions, such as those of James Croll, originating in 1864.

The pace of palaeoclimatic research has accelerated over recent decades. Quantitative and well-dated records of climate fluctuations over the last 100 kyr have brought a more comprehensive view of how climate changes occur, as well as the means to test elements of the astronomical theory. By the 1950s, studies of deep-sea cores suggested that the ocean temperatures may have been different during glacial times. Ewing and Donn proposed that changes in ocean circulation actually could initiate an ice age. In the 1960s, the works of Emiliani and Shackleton showed the potential of isotopic measurements in deepsea sediments to help explain Quaternary changes. In the 1970s, it became possible to analyse a deep-sea core time series of more than 700 kyr, thereby using the last reversal of the Earth's magnetic field to establish a dated chronology. This deep-sea observational record clearly showed the same periodicities found in the astronomical forcing, immediately providing strong support to Milankovitch's theory.

Ice cores provide key information about past climates, including surface temperatures and atmospheric chemical composition. The bubbles sealed in the ice are the only available samples of these past atmospheres. The first deep ice cores from Vostok in Antarctica provided additional evidence of the role of astronomical forcing. They also revealed a highly correlated evolution of temperature changes and atmospheric composition, which was subsequently confirmed over the past 400 kyr and now extends to almost 1 Myr. This discovery drove research to understand the causal links between greenhouse gases and climate change.

The same data that confirmed the astronomical theory also revealed its limits: a linear response of the climate system to astronomical forcing could not explain entirely the observed fluctuations of rapid ice-age terminations preceded by longer cycles of glaciations. The importance of other sources of climate variability was heightened by the discovery of abrupt climate changes. In this context, 'abrupt' designates regional events of large amplitude, typically a few degrees celsius, which occurred within several decades–much shorter than the thousand-year time scales that characterise changes in astronomical forcing. Abrupt temperature changes were first revealed by the analysis of deep ice cores from Greenland.

Oeschger et al. recognised that the abrupt changes during the termination of the last ice age correlated with cooling in Gerzensee and suggested that regime shifts in the Atlantic Ocean circulation were causing these widespread changes. The synthesis of palaeoclimatic observations by Broecker and Denton invigourated the community over the next decade. By the end of the 1990s, it became clear that the abrupt climate changes during the last ice age, particularly in the North Atlantic regions as found in the Greenland ice cores, were numerous, indeed abrupt and of large amplitude. They are now referred to as Dansgaard-Oeschger events. A similar variability is seen in the North Atlantic Ocean, with north-south oscillations of the polar front and associated changes in ocean temperature and salinity. With no obvious external forcing, these changes are thought to be manifestations of the internal variability of the climate system.

The importance of internal variability and processes was reinforced in the early 1990s with analysis of records with high temporal resolution. New ice cores new ocean cores from regions with high sedimentation rates, as well as lacustrine sediments and cave stalagmites produced additional evidence for unforced climate changes, and revealed a large number of abrupt changes in many regions throughout the last glacial cycle. Long sediment cores from the deep ocean were used to reconstruct the thermohaline circulation connecting deep and surface waters and to demonstrate the participation of the ocean in these abrupt climate changes during glacial periods.

By the end of the 1990s, palaeoclimate proxies for a range of climate observations had expanded greatly. The analysis of deep corals provided indicators for nutrient content and mass exchange from the surface to deep water, showing abrupt variations characterised by synchronous changes in surface and deep-water properties. Precise measurements of the CH_4 abundances in polar ice cores showed that they changed in concert with the Dansgaard-Oeschger events and thus allowed for synchronisation of the dating across ice cores. The characteristics of the antarctic temperature variations and their relation to the Dansgaard-Oeschger events in Greenland were consistent with the simple concept of a bipolar seesaw caused by changes in the thermohaline circulation of the Atlantic Ocean. This work underlined the role of the ocean in transmitting the signals of abrupt climate change.

Abrupt changes are often regional, for example, severe droughts lasting for many years have changed civilizations, and have occurred during the last 10 kyr of stable warm climate. This result has altered the notion of a stable climate during warm epochs, as previously suggested by the polar ice cores. The emerging picture of an unstable oceanatmosphere system has opened the debate of whether human interference through greenhouse gases and aerosols could trigger such events. Palaeoclimate reconstructions cited in the FAR were based on various data, including pollen records, insect and animal remains, oxygen isotopes and other geological data from lake varves, loess, ocean sediments, ice cores and glacier termini.

These records provided estimates of climate variability on time scales up to millions of years. A climate proxy is a local quantitative record that is interpreted as a climate variable using a transfer function that is based on physical principles and recently observed correlations between the two records. The combination of instrumental and proxy data began in the 1960s with the investigation of the influence of climate on the proxy data, including tree rings, corals and ice cores. Phenological and historical data are also a valuable source of climatic reconstruction for the period before instrumental records became available. Such documentary data also need calibration against instrumental data to extend and reconstruct the instrumental record. With the development of multi-proxy reconstructions, the climate data were extended not only from local to global, but also from instrumental data to patterns of climate variability. Most of these reconstructions were at single sites and only loose efforts had been made to consolidate records.

Mann et al. made a notable advance in the use of proxy data by ensuring that the dating of different records lined up. Thus, the true spatial patterns of temperature variability and change could be derived, and estimates of NH average surface temperatures were obtained. The Working Group I WGI FAR noted that past climates could provide analogues. Fifteen years of research since that assessment has identified a range of variations and instabilities in the climate system that occurred during the last 2 Myr of glacial-interglacial cycles and in the super-warm period of 50 Ma. These past climates do not appear to be analogues of the immediate future, yet they do reveal a wide range of climate processes that need to be understood when projecting 21st century climate change.

BIOGEOCHEMISTRY AND RADIATIVE FORCING

The modern scientific understanding of the complex and interconnected roles of greenhouse gases and aerosols in climate change has undergone rapid evolution over the last two decades. While the concepts were recognised and outlined in the 1970s, the publication of generally accepted quantitative results coincides with, and was driven in part by, the questions asked by the IPCC beginning in 1988. Thus, it is instructive to view the evolution of this topic as it has been treated in the successive IPCC reports. The WGI FAR codified the key physical and biogeochemical processes in the Earth system that relate a changing climate to atmospheric composition, chemistry, the carbon cycle and natural ecosystems. The science of the time, as summarised in the FAR, made a clear case for anthropogenic interference with the climate system.

In terms of greenhouse agents, the main conclusions from the WGI FAR Policymakers Summary are still valid today:

- 'Emissions resulting from human activities are substantially increasing the atmospheric concentrations of the greenhouse gases: CO_2, CH_4, CFCs, N_2O';
- 'Some gases are potentially more effective';
- Feedbacks between the carbon cycle, ecosystems and atmospheric greenhouse gases in a warmer world will affect CO2 abundances; and
- GWPs provide a metric for comparing the climatic impact of different greenhouse gases, one that integrates both the radiative influence and biogeochemical cycles.

The climatic importance of tropospheric ozone, sulphate aerosols and atmospheric chemical feedbacks were proposed by scientists at the time and noted in the assessment. For example, early global chemical modelling results argued that global tropospheric ozone, a greenhouse gas, was controlled by emissions of the highly reactive gases nitrogen oxides, carbon monoxide and non-methane hydrocarbons. In terms of sulphate aerosols, both the direct radiative effects and the indirect effects on clouds were acknowledged, but the importance of carbonaceous aerosols from fossil fuel and biomass combustion was not recognised.

The concept of radiative forcing as the radiative imbalance in the climate system at the top of the atmosphere caused by the addition of a greenhouse gas was established at the time and summarised in the WGI FAR. Agents of RF included the direct greenhouse gases, solar radiation, aerosols and the Earth's surface albedo. What was new and only briefly mentioned was that 'many gases produce indirect effects on the global radiative forcing'. The innovative global modelling work of Derwent showed that emissions of the reactive but non-greenhouse gases–NOx, CO and NMHCs–altered atmospheric chemistry and thus changed the abundance of other greenhouse gases. Indirect GWPs for NOx, CO and VOCs were proposed.

The projected chemical feedbacks were limited to short-lived increases in tropospheric ozone. By 1990, it was clear that the RF from tropospheric ozone had increased over the 20th century and stratospheric ozone had decreased since 1980 but the associated RFs were not evaluated in the assessments. Neither was the effect of anthropogenic sulphate aerosols, except to note in the FAR that 'it is conceivable that this radiative forcing has been of a comparable magnitude, but of opposite sign, to the greenhouse forcing earlier in the century'. Reflecting in general the community's concerns about this relatively new measure of climate forcing, RF bar charts appear only in the underlying FAR stages, but not in the FAR Summary. Only the long-lived greenhouse gases are shown, although sulphate aerosols direct effect in the future is noted with a question mark.

The cases for more complex chemical and aerosol effects were becoming clear, but the scientific community was unable at the time to reach general agreement on the existence, scale and magnitude of these indirect effects. Nevertheless, these early discoveries drove the research agendas in the early 1990s. The widespread development and application of global chemistrytransport models had just begun with international workshops. In the Supplementary Report to the FAR, the indirect chemical effects of CO, NOx and VOC were reaffirmed, and the feedback effect of CH4 on the tropospheric hydroxyl radical was noted. Aerosolclimate interactions still focused on sulphates, and the assessment of their direct RF for the NH was now somewhat quantitative as compared to the FAR. Stratospheric ozone depletion was noted as causing a significant and negative RF, but not quantified. Ecosystems research at this time was identifying the responses to climate change and CO_2 increases, as well as altered CH_4 and N_2O fluxes from natural systems; however, in terms of a community assessment it remained qualitative.

By 1994, with work on SAR progressing, the Special Report on Radiative Forcing reported significant breakthroughs in a set of stages limited to assessment of the carbon cycle, atmospheric chemistry, aerosols and RF. The carbon budget for the 1980s was analysed not only from bottomup emissions estimates, but also from a top-down approach including carbon isotopes. A first carbon cycle assessment was performed through an international model and analysis workshop examining terrestrial and oceanic uptake to better quantify the relationship between CO2 emissions and the resulting increase in atmospheric abundance. Similarly, expanded analyses of the global budgets of trace gases and aerosols from both natural and anthropogenic sources highlighted the rapid expansion of biogeochemical research.

The first RF bar chart appears, comparing all the major components of RF change from the pre-industrial period to the present. Anthropogenic soot aerosol, with a positive RF, was not in the 1995 Special Report but was added to the SAR. In terms of atmospheric chemistry, the first open-invitation modelling study for the IPCC recruited 21 atmospheric chemistry models to participate in a controlled study of photochemistry and chemical feedbacks. These studies demonstrated a robust consensus about some indirect effects, such as the CH_4 impact on atmospheric chemistry, but great uncertainty about others, such as the prediction of tropospheric ozone changes.

The model studies plus the theory of chemical feedbacks in the CH_4-CO-OH system firmly established that the atmospheric lifetime of a perturbation of CH_4 emissions was about 50 per cent greater than reported in the FAR. There was still no consensus on quantifying the past or future changes in tropospheric ozone or OH. In the early 1990s, research on aerosols as climate forcing agents expanded. Based on new research, the range of climaterelevant aerosols was extended for the first time beyond sulphates to include nitrates, organics, soot, mineral dust and sea salt.

Quantitative estimates of sulphate aerosol indirect effects on cloud properties and hence RF were sufficiently well established to be included in assessments, and carbonaceous aerosols from biomass burning were recognised as being comparable in importance to sulphate. Ranges are given in the special report for direct sulphate RF and biomass-burning aerosols.

The aerosol indirect RF was estimated to be about equal to the direct RF, but with larger uncertainty. The injection of stratospheric aerosols from the eruption of Mt. Pinatubo was noted as the first modern test of a known radiative forcing, and indeed one climate model accurately predicted the temperature response. In the one-year interval between the special report and the SAR, the scientific understanding of aerosols grew. The direct anthropogenic aerosol forcing was reduced to –0.5 W m–2. The RF bar chart was now broken into aerosol components with a separate range for indirect effects. Throughout the 1990s, there were concerted research programmes in the USA and EU to evaluate the global environmental impacts of aviation. Several national assessments culminated in the IPCC Special Report on Aviation and the Global Atmosphere which assessed the impacts on climate and global air quality.

An open invitation for atmospheric model participation resulted in community participation and a consensus on many of the environmental impacts of aviation. The direct RF of sulphate and of soot aerosols was likewise quantified along with that of contrails, but the impact on cirrus clouds that are sometimes generated downwind of contrails was not. The assessment re- affirmed that RF was a first-order metric for the global mean surface temperature response, but noted that it was inadequate for regional climate change, especially in view of the largely regional forcing from aerosols and tropospheric ozone. By the end of the 1990s, research on atmospheric composition and climate forcing had made many important advances. The TAR was able to provide a more quantitative evaluation in some areas. For example, a large, open-invitation modelling workshop was held for both aerosols and tropospheric ozone-OH chemistry.

This workshop brought together as collaborating authors most of the international scientific community involved in developing and testing global models of atmospheric composition. In terms of atmospheric chemistry, a strong consensus was reached for the first time that science could predict the changes in tropospheric ozone in response to scenarios for CH4 and the indirect greenhouse gases and that a quantitative GWP for CO could be reported.

Further, combining these models with observational analysis, an estimate of the change in tropospheric ozone since the pre-industrial era–with uncertainties–was reported. The aerosol workshop made similar advances in evaluating the impact of different aerosol types. There were many different representations of uncertainty in the TAR, and the consensus RF bar chart did not generate a total RF or uncertainties for use in the subsequent IPCC Synthesis Report.

CLIMATE CHANGE MITIGATION

Climate change mitigation is action to decrease the intensity of radiative forcing in order to reduce the potential effects of global warming. Mitigation is distinguished from adaptation to global warming, which involves acting to tolerate the effects of global warming. Most often, climate change mitigation scenarios involve reductions in the concentrations of greenhouse gases, either by reducing their sources or by increasing their sinks.

Scientific consensus on global warming, together with the precautionary principle and the fear of abrupt climate change is leading to increased effort to develop new technologies and sciences and carefully manage others in an attempt to mitigate global warming. Most means of mitigation appear effective only for preventing further warming, not at reversing existing warming. The Stern Review identifies several ways of mitigating climate change. These include reducing demand for emissions-intensive goods and services, increasing efficiency gains, increasing use and development of low-carbon technologies, and reducing fossil fuel emissions.

The energy policy of the European Union has set a target of limiting the global temperature rise to 2°C compared to preindustrial levels, of which 0.8°C has already taken place and another 0.5–0.7°C is already committed. The 2°C rise is typically associated in climate models with a carbon dioxide- equivalent concentration of 400–500 ppm by volume; the current level of carbon dioxide alone is 383 ppm by volume, and rising at 2 ppm annually.

Hence, to avoid a very likely breach of the 2°C target, CO_2 levels would have to be stabilised very soon; this is generally regarded as unlikely, based on current programmes in place to date. The importance of change is showed by the fact that world economic energy efficiency is presently improving at only half the rate of world economic growth.

CLIMATE CHANGE AND ECOSYSTEMS

Unchecked global warming could affect most terrestrial ecoregions. Increasing global temperature means that ecosystems will change; some species are being forced out of their habitats because of changing conditions, while others are flourishing. Secondary effects of global warming, such as lessened snow cover, rising sea levels, and weather changes, may influence not only human activities but also the ecosystem.

For the IPCC Fourth Assessment Report, experts assessed the literature on the impacts of climate change on ecosystems. Rosenzweig et al. concluded that over the last three decades, human-induced warming had likely had a discernable influence on many physical and biological systems. Schneider et al. concluded, with very high confidence, that regional temperature trends had already affected species and ecosystems around the world. With high confidence, they concluded that climate change would result in the extinction of many species and a reduction in the diversity of ecosystems.

- Terrestrial ecosystems and biodiversity: With a warming of 3°C, relative to 1990 levels, it is likely that global terrestrial vegetation would become a net source of carbon. With high confidence, Schneider et al. concluded that a global mean temperature increase of around 4°C by 2100 would lead to major extinctions around the globe.
- Marine ecosystems and biodiversity: With very high confidence, Schneider et al. concluded that a warming of 2°C above 1990 levels would result in mass mortality of coral reefs globally.
- Freshwater ecosystems: Above about a 4°C increase in global mean temperature by 2100, Schneider et al. concluded, with high confidence, that many freshwater species would become extinct.

Studying the association between Earth climate and extinctions over the past 520 million years, scientists from the University of York write, "The global temperatures predicted for the coming centuries may trigger a new 'mass extinction event', where over 50 per cent of animal and plant species would be wiped out." Many of the species at risk are Arctic and Antarctic fauna such as polar bears and Emperor Penguins. In the Arctic, the waters of Hudson Bay are ice-free for three weeks longer than they were thirty years ago, affecting polar bears, which prefer to hunt on sea ice. Species that rely on cold weather conditions such as gyrfalcons, and Snowy Owls that prey on lemmings that use the cold winter to their advantage may be hit hard. Marine invertebrates enjoy peak growth at the temperatures they have adapted to, regardless of how cold these may be, and cold-blooded animals found at greater latitudes and altitudes generally grow faster to compensate for the short growing season.

Warmer-than-ideal conditions result in higher metabolism and consequent reductions in body size despite increased foraging, which in turn elevates the risk of predation. Indeed, even a slight increase in temperature during development impairs growth efficiency and survival rate in rainbow trout. Rising temperatures are beginning to have a noticeable impact on birds, and butterflies have shifted their ranges northward by 200 km in Europe and North America. Plants lag behind, and larger animals' migration is slowed down by cities and roads. In Britain, spring butterflies are appearing an average of 6 days earlier than two decades ago.

A 2002 article in Nature surveyed the scientific literature to find recent changes in range or seasonal behaviour by plant and animal species. Of species showing recent change, 4 out of 5 shifted their ranges towards the poles or higher altitudes, creating "refugee species". Frogs were breeding, flowers blossoming and birds migrating an average 2.3 days earlier each decade; butterflies, birds and plants moving towards the poles by 6.1 km per decade. A 2005 study concludes human activity is the cause of the temperature rise and resultant changing species behaviour, and links these effects with the predictions of climate models to provide validation for them. Scientists have observed that Antarctic hair grass is colonizing areas of Antarctica where previously their survival range was limited.

Mechanistic studies have documented extinctions due to recent climate change: McLaughlin et al. documented two populations of Bay checkerspot butterfly being threatened by precipitation change. Parmesan states, "Few studies have been conducted at a scale that encompasses an entire species" and McLaughlin et al. agreed "few mechanistic studies have linked extinctions to recent climate change." Daniel Botkin and other authors in one study believe that projected rates of extinction are overestimated.

Many species of freshwater and saltwater plants and animals are dependent on glacier-fed waters to ensure a cold water habitat that they have adapted to. Some species of freshwater fish need cold water to survive and to reproduce, and this is especially true with Salmon and Cutthroat trout. Reduced glacier run-off can lead to insufficient stream flow to allow these species to thrive. Ocean krill, a cornerstone species, prefer cold water and are the primary food source for aquatic mammals such as the Blue Whale. Alterations to the ocean currents, due to increased freshwater inputs from glacier melt, and the potential alterations to thermohaline circulation of the worlds oceans, may affect existing fisheries upon which humans depend as well.

Forests

Pine forests in British Columbia have been devastated by a pine beetle infestation, which has expanded unhindered since 1998 at least in part due to the lack of severe winters since that time; a few days of extreme cold kill most mountain pine beetles and have kept outbreaks in the past naturally contained. The infestation, which has killed about half of the province's lodgepole pines is an order of magnitude larger than any previously recorded outbreak and passed via unusually strong winds in 2007 over the continental divide to Alberta. An epidemic also started, be it at a lower rate, in 1999 in Colourado, Wyoming, and Montana. The United States forest service predicts that between 2011 and 2013 virtually all 5 million acres of Colourado's lodgepole pine trees over five inches in diametre will be lost.

As the northern forests are a carbon sink, while dead forests are a major carbon source, the loss of such large areas of forest has a positive feedback on global warming. In the worst years, the carbon emission due to beetle infestation of forests in British Columbia alone approaches that of an average year of forest fires in all of Canada or five years worth of emissions from that country's transportation sources.

Besides the immediate ecological and economic impact, the huge dead forests provide a fire risk. Even many healthy forests appear to face an increased risk of forest fires because of warming climates. The 10-year average of boreal forest burned in North America, after several decades of around 10,000 km^2 has increased steadily since 1970 to more than 28,000 km^2 annually. Though this change may be due in part to changes in forest management practices, in the western U.S., since 1986, longer, warmer summers have resulted in a fourfold increase of major wildfires and a sixfold increase in the area of forest burned, compared to the period from 1970 to 1986. A similar increase in wildfire activity has been reported in Canada from 1920 to 1999.

Forest fires in Indonesia have dramatically increased since 1997 as well. These fires are often actively started to clear forest for agriculture. They can set fire to the large peat bogs in the region and the COreleased by these peat bog fires has been estimated, in an average year, to be 15 per cent of the quantity of Co-produced by fossil fuel combustion.

Mountains

Mountains cover approximately 25 per cent of earth's surface and provide a home to more than one-tenth of global human population. Changes in global climate pose a number of potential risks to mountain habitats. Researchers expect that over time, climate change will affect mountain and lowland ecosystems, the frequency and intensity of forest fires, the diversity of wildlife, and the distribution of water.

Studies suggest that a warmer climate in the United States would cause lower-elevation habitats to expand into the higher alpine zone. Such a shift would encroach on the rare alpine meadows and other high-altitude habitats. High-elevation plants and animals have limited space available for new habitat as they move higher on the mountains in order to adapt to long-term changes in regional climate.

Changes in climate will also affect the depth of the mountains snowpacks and glaciers. Any changes in their seasonal melting can have powerful impacts on areas that rely on freshwater run-off from mountains. Rising temperature may cause snow to melt earlier and faster in the spring and shift the timing and distribution of run-off. These changes could affect the availability of freshwater for natural systems and human uses.

Ecological Productivity

- A document by Smith and Hitz, it is reasonable to assume that the relationship between increased global mean temperature and ecosystem productivity is parabolic. Higher carbon dioxide concentrations will favourably affect plant growth and demand for water. Higher temperatures could initially be favourable for plant growth. Eventually, increased growth would peak then decline.
- The IPCC, a global average temperature increase exceeding 1.5–2.5°C would likely have a predominantly negative impact on ecosystem goods and services, e.g., water and food supply.
- Research done by the Swiss Canopy Crane Project suggests that slow-growing trees only are stimulated in growth for a short period under higher CO_2 levels, while faster growing plants like liana benefit in the long term. In general, but especially in rainforests, this means that liana become the prevalent species; and because they decompose much faster than trees their carbon content is more quickly returned to the atmosphere. Slow growing trees incorporate atmospheric carbon for decades.

TIME SCALE OF CLIMATIC CHANGE

The importance of considering different time scales when investigating climate change has already been identified. Climate varies on all time scales, in response to random and periodic forcing factors. Across all time periods from a few years to hundreds of millions of years there is a white (background) noise of random variations of the climate, caused by internal processes and associated feedback mechanisms, often referred to as stochastic or random mechanisms. Such randomness accounts for much of the climate variation, and owes its existence to the complex and chaotic behaviour of the climate system in responding to forcing. An essential corollary of the existence of random processes is that a large proportion of climate variation cannot be predicted.

Of far more relevance are the periodic forcing factors, for by understanding their mechanisms and the impacts they have on the global climate, it is possible to predict future climate change. How the climate system responds to periodic forcing factors, however, is often not clear. If it is assumed that the climate system responds in a linear fashion to periodic forcing, variations in climate will exhibit similar periodicity. If, however, the response of the system to forcing is strongly non-linear, the periodicities in the response will not necessarily be identical to the periodicities in the forcing factor(s). Frequently, the climate responds in a fashion intermediate between the two.

There are many climate forcing factors spanning an enormous range of periodicities. The longest, 200 to 500 million years, involves the passage of our Solar System through the galaxy, and the variations in galactic dust. These may be considered to be external forcing mechanisms. Other long time scale variations (10^6 to 10^8 years) include the non-radiative forcing mechanisms, such as continental drift, orogeny (mountain building) and isostasy (vertical movements in the Earth's crust affecting sea level). The response of the climate system to this combination of forcing factors itself depends upon the different response times of the various components of the system. The overall climatic response will then be determined by the interactions between the components. The atmosphere, surface snow and ice, and surface vegetation typically respond to climatic forcing over a period of hours to days. The surface ocean has a response time measured in years, whilst the deep ocean and mountain glaciers vary only over a period spanning hundreds of years. Large ice sheets advance and withdraw over thousands of years whilst parts of the geosphere (e.g. continental weathering of rocks) respond only to forcing periods lasting hundreds of thousands to millions of years.

The response of the climate system to episodes of forcing can be viewed as a form of resonance. When the time period of forcing matches most closely the response time of a particular system component, the climatic response will be greatest within that component.

Milankovitch forcing, for example, with periods of tens of thousands of years will be manifest in the response of the ice sheets, and the overall response of the climate system will be dominated by changes within the cryosphere. In addition, longer response times of certain components of the climate system modulate, through feedback processes, the short term responses. The response of the deep ocean to short term forcing (e.g. enhanced greenhouse effect, solar variations, for example, will tend to attenuate or smooth the response of the atmosphere.

Throughout the remainder, it should be recognised that a range of time scales applies to climate forcing mechanisms, radiative and non-radiative, external and internal, and to the response of the different components of the climate system.

Climate Sensitivity

The concept of feedback is related to the climate sensitivity or climate stability. It is useful to have a measure of the strength of various feedback processes which determine the ultimate response of the climate system to any change in radiative forcing. In general terms, an initial change in temperature due to a change in radiative forcing, $\Delta T_{forcing}$, is modified by the complex combination of feedback processes such that:

$$\Delta T_{final} = \Delta T_{forcing} + \Delta T_{feedback}$$

where $\Delta T_{feedback}$ is the temperature change resulting from feedback and ΔT_{final} is the overall change in temperature between the initial and final equilibrium states. The degree to which feedback processes influence the final climatic response is a measure of the sensitivity of the climate system.

Equation can be rewritten as follows:

$$\Delta T_{final} = f\Delta T_{forcing}$$

where *f* is called the feedback factor. When only one feedback mechanism is operative the solution to equation 6 is simple, assuming both *f* and $\Delta T_{forcing}$ are known. When more than one feedback is operative, matters become more complicated. For two feedbacks, the net effect is given by:

$$f = f_1 f2 / (f_1 + f2 - f_1 f2)$$

where f_1 and f2 are the feedback factors of the two feedback processes. Clearly, the feedback factors are neither additive nor multiplicative. A feedback operating alone with a factor of 2 would double the initial climatic response to forcing. If a second feedback with factor 1.5 acted with it, the overall feedback would be enhanced by a factor of 6. It can be seen, then, that a combination of feedback processes could dramatically affect the climate as it responds to only a small change in radiative forcing.

The climate's sensitivity can be mathematically determined in another way. From satellite observations, it has been shown that changes in global temperature are approximately proportional to changes in radiative forcing. If we assume an instantaneous † change in climate, from one equilibrium state to another, then:

$$\Delta Q = \lambda \Delta T$$

where ΔQ is change in radiative forcing (expressed in terms of the net downward radiative flux at the top of the troposphere), ΔT is the change in global temperature, and λ is a measure of the climate sensitivity.

Based on equation, the climate sensitivity is usually expressed in terms of the temperature change associated with a specified change in radiative forcing, usually a doubling of the atmospheric carbon dioxide content. Thus, the carbon dioxide doubling temperature, ΔT_{2x}, is given by,

$$\Delta T_{2x} = \Delta Q_{2x} / \lambda$$

where ΔQ_{2x} is 4.2 Wm^{-2}. The magnitude (and sign) of ΔT_{2x} will depend on λ, the climate sensitivity, which is determined by the net affect of the climate feedback processes. Despite extensive climate modelling over the last 2 decades to understand the problem of contemporary global, it is this parametre that is proving hard to define numerically.

As explained earlier, the idea of static equilibrium and instantaneous climatic response represents an unrealistic situation. To take into account the dynamic and transient nature of the climate's response to forcing, a more complicated equation linking ΔT and ΔQ is required if the evolution of the response with time is to be determined.

In the case:

$$\Delta Q = \lambda \Delta T + C d\Delta T / \delta t$$

Here, the change in radiative forcing, DQ, is balanced by:

- The change in the outgoing radiative flux at the tropopause caused by the response of the climate system including feedback; and
- The energy stored in the system, $C\delta\Delta T/\delta t$, where C is the system's heat capacity and *t* is time.

The latter term in equation simulates the time-dependent nature of the climate system's response. The main contributor to the system's heat capacity is the world's oceans. The heat capacity of water is large compared to that of air and is therefore able to store much more energy. Additionally, the high heat capacity means that the oceans take time to heat up (or cool down) and hence slow the surface temperature response to any change in radiative forcing: the transient response will always be less than the equilibrium response.

Solution of Equation leads to the definition of the response time of the climate system, τ, such that:

$$\tau = C / \lambda$$

If the heat capacity of the climate system is large, the response time is large. Equally, if the climate sensitivity is small the response time is large. λ is also known as the radiative damping coefficient. Here, analogy is made between the response of the climate system and an oscillating spring. If a spring has a high damping coefficient, it will stop oscillating soon after it has been set in motion. Similarly, if the radiative damping coefficient is large, the climate will respond quickly and τ will be small.

THE EFFECTS OF CLIMATE CHANGE ON ECOSYSTEMS

One way that scientists gain understanding of how global warming will affect ecosystems is to analyse the effects of past climate on paleoecosystems. A paleoecosystem is an ecosystem that existed in a former geologic time period. By relating vegetative cover to past climates, models can be developed. Once a reliable model is created, input variables such as CO_2 can be varied and the results analysed. An example of what the effect would be like on populations of Douglas fir in the northwestern United States if the CO_2 content in the atmosphere were double what it was before the Industrial Revolution was modeled by the U.S. Geological Survey and is shown in the illustration that follows. In a study conducted by the U.S. Global Change Research Programme in Washington, D.C., first in 2001, then updated in 2004, in trying to predict what the effects of future climate change would have on ecosystems, they concluded that "climate change has the potential to affect the structure, function, and regional distribution of ecosystems, and thereby affect the goods and services they provide." They based their conclusions on a modeling and analysis project they conducted called the Vegetation/Ecosystem Modeling and Analysis Project.

This project was used to generate future ecosystem scenarios for the conterminous United States based on model-simulated responses to both the Canadian and Hadley scenarios of climate change. The VEMAP was subsequently used in a validation exercise for a Dynamic Global Vegetation Model by Oregon State University and the U.S. Forest Service in 2008. Their MC1-DGVM was used as the input data in both the VEMAP and VINCERA. Their MC1 was run on both the VEMAP and VINCERA climate and soil input data to document how a change in the inputs can affect model outcome. The simulation results under the two sets of future climate scenarios were compared to see how different inputs can affect vegetation distribution and carbon budget projections.

The results indicated that "under all future scenarios, the interior west of the United States becomes woodier as warmer temperatures and available moisture allow trees to get established in grasslands areas. Concurrently, warmer and drier weather causes the eastern deciduous and mixed forests to shift to a more open canopy woodland or savannatype while boreal forests disappear almost entirely from the Great Lakes area by the end of the 21st century. While under VEMAP scenarios the model simulated large increases in carbon storage in a future woodier west, the drier VINCERA scenarios accounted for large carbon losses in the east and only moderate gains in the west. But under all future climate scenarios, the total area burned by wildfires increased." The similarities of the two models served to validate the VEMAP project.

The Hadley model was developed by the Hadley Centre for Climate Prediction and Research in England. Also referred to as the Met Office Hadley Centre for Climate Change, it is based at the headquarters of the Met Office in Exeter. It is the key institution in the United Kingdom for climate research. It is currently involved not only with understanding the physical, chemical, and biological processes within the climate system, but also with developing working models to explain current phenomena and to predict future climate change. It also monitors global and national climate variability and change and strives to determine the causes of the fluctuations. The Canadian climate model was developed by the Canadian Centre for Climate Modelling and Analysis. The CCCma is a division of the Climate Research Branch of Environment Canada based out of the University of Victoria, Victoria, British Columbia.

Its specific focus is on climate change and modeling. In the past nine years, the CCCma has produced three atmospheric and three atmospheric/oceanic general circulation models, making them one of the international leaders in climate change research. What they found was that over the next few decades climate change in the United States will most likely lead to increased plant productivity as a result of increasing levels of CO_2 in the atmosphere. There will also be an increase in terrestrial carbon storage for many parts of the country, especially the areas that become warmer and wetter.

The southeast will most likely see reduced productivity and, therefore, a decrease in carbon storage. By the end of the 21st century, many areas of the country will have experienced changes in the distribution of vegetation. Wetter areas will see the growth of more trees; drier areas will have drier soils that will cause forested areas to die off and be replaced by savanna/grassland ecosystems.

Modeling the vegetation evolution and adaptation is more difficult. The study focused on two time periods: 2025-2034 and 2090-2099. In the near term, biogeochemical changes are expected to dominate the ecological responses. Biogeochemical responses include changes based on the natural cycles of carbon, nutrients and water. The responses are affected by changing environmental conditions such as temperature, precipitation, solar radiation, soil texture, and atmospheric CO_2.

It is these natural cycles that affect carbon capture by plants with photosynthesis, soil nitrogen processes, and water transfer. These biogeochemical factors are what influence the production of vegetation. In the results from the near-term biogeochemical model, the scientists concluded that there would be an increase in CO_2. They estimate that currently the average carbon storage rate is 66/Tg/yr. The Hadley model predicted that carbon storage rates by 2025-2034 would increase to 117/ Tg/yr. The Canadian model estimated CO_2 to increase 96 Tg/yr. The Canadian model projected that the southeastern ecosystems will lose carbon in the near term because they predict the climate there will become hot and dry. The biogeography models look at the changing landscape based on changes in CO_2, evapotranspiration, vegetation establishment, and competition between species, growth rates, and life cycle/mortality rates. In this model, scientists at both the Hadley and Canadian Centre for Climate Change agree that vegetation will be able to freely move from one location to another.

Changes in vegetation distribution will vary from region to region as follows:

- *Northeast:* Forests remain the dominant natural vegetation, but forest mixes will change. There will also be some increase in savannas and wetlands.
- *Southeast:* Forests remain the dominant ecosystem, but mixes change. Savannas and grasslands encroach on forests, especially towards the end of the 21st century. Drought and wildfires contribute significantly to forest destruction.
- *Midwest:* Forests remain the dominant land cover, but changes in species type occur. There will be a modest expansion of savannas and grasslands.
- *Great Plains:* Slight increase in woody vegetation.
- *West:* The areas of desert ecosystems shrink, and forest ecosystems grow.
- *Northwest:* Forested areas grow slightly.

A separate study conducted by the Canadian government predicts the following ecosystem changes as a result of changing climate over the next 100 years:

- Coastlines:
  - Flooding and erosion in coastal regions
  - Sea-level rise
- Forests:
  - Increase in pests
  - Increased levels of drought and wildfire
- Plants and animals:
  - Warmer temperatures could make water supplies more scarce, having a negative impact on plants and animals, not giving them time to adjust.
- Crops:
  - In some areas, warmer climate may allow a three- to five-week extension of the frost-free season, which could benefit

commercial agriculture.
- In other areas, drier soils and lack of water will have a negative impact on agricultural productivity.

- Wells:
  - The quality and quantity of drinking water may be threatened by increasing drought.
- Harsh weather:
  - Winter storms, floods, drought, heat waves, and tornadoes could become more frequent and severe.
- Fisheries:
  - Populations and ranges of species sensitive to changes in water temperature will be negatively affected.
  - Salmon harvests will be lower in the Pacific.
  - Changes in ocean currents may have a negative impact on the fisheries in the Atlantic.
- Lakes and rivers:
  - Water levels will decline under the influence of drought, negatively affecting drinking water quality.
  - Use of lakes for transportation, recreation, and fishing, and the ability to generate electricity may be curtailed under droughtlike conditions.
  - Other areas that may have an increase in precipitation may experience flooding, rising sea levels, and severe storms.

In February 2005, the Met Office in Exeter, England, issued a report titled "Avoiding Dangerous Climate Change." The objective of the study was to determine what levels of CO_2 were considered the tipping point for dangerous climate change with harmful effects on ecosystems and what actions could be taken now to avoid such an outcome. In the report, then prime minister *Tony Blair stated:* "It is now plain that the emission of greenhouse gases... is causing global warming at a rate that is unsustainable." Environment Secretary Margaret Beckett stated," The report's conclusions would be a shock to many people.

The thing that is perhaps not so familiar to members of the public... is this notion that we could come to a tipping point where change could be irreversible. We're not talking about it happening over five minutes, of course, maybe over a thousand years, but it's the irreversibility that I think brings it home to people." The report, published by the British government, says there is only a small chance of greenhouse gas emissions being kept below"dangerous" levels. It warns that the Greenland ice sheet could melt, causing sea levels to rise by 23 feet over the next 1,000 years. It also warns that developing countries will be the hardest hit. The report also states, based on the vulnerability of many of the world's ecosystems, that the

European Union (EU) has adopted a target of preventing an increase in global temperature of more than 3.3°F (2°C). Some believe even that may be too high. The report states: "Above two degrees, the risks increase very substantially," with "potentially large numbers of extinctions" and "major increases in hunger and water shortage risks... particularly in developing countries." In order to meet their goals, British scientists have advised that CO_2 levels should be stabilized at 450 parts per million (ppm) or below. Currently the atmosphere contains 380 ppm. In response to this, the British government's chief scientific adviser, Sir David King, said that was unlikely to happen. He stated,"We're going to be at 400 ppm in 10 years' time. I predict that without any delight in saying it." Myles Allen, an expert on atmospheric physics at Oxford University, said that: "Assessing a 'safe level' of carbon dioxide in the atmosphere was 'a bit like asking a doctor what's a safe number of cigarettes to smoke per day."

The report does conclude, however, that there are technological options available to reduce CO2 emissions that will need to be used. The study also concluded that the biggest obstacles involved with using these new technologies, along with renewable resources of energy and"clean coal," are the current economic investments and traditionally strong bond to the oil industry, cultural attitudes that oppose change, and simple lack of awareness by many people. Various conservation organizations currently involved in the battle against global warming, such as the Union of Concerned Scientists (UCS), the Defenders of Wildlife, and the World Wildlife Fund (WWF), also support these ideas.

IMPACT OF CLIMATE CHANGE ON AGRICULTURE

Despite technological advances, such as improved varieties, genetically modified organisms, and irrigation systems, weather is still a key factor in agricultural productivity, as well as soil properties and natural communities. The effect of climate on agriculture is related to variabilities in local climates rather than in global climate patterns. The Earth's average surface temperature has increased by 1 degree F in just over the last century. Consequently, agronomists consider any assessment has to be individually consider each local area. On the other hand, agricultural trade has grown in recent years, and now provides significant amounts of food, on a national level to major importing countries, as well as comfortable income to exporting ones. The international aspect of trade and security in terms of food implies the need to also consider the effects of climate change on a global scale.

A study published in Science suggests that, due to climate change, "southern Africa could lose more than 30 per cent of its main crop, maize, by 2030. In South Asia losses of many regional staples, such as rice, millet and maize could top 10 per cent". The 2001 IPCC Third Assessment Report concluded that the poorest countries would be hardest hit, with reductions in crop yields in most tropical and sub-tropical regions due to decreased water availability, and new or changed insect pest incidence. In Africa and Latin America many rainfed crops are near their maximum temperature tolerance, so that yields are likely to fall sharply for even small climate changes; falls in agricultural productivity of up to 30 per cent over the 21st century are projected. Marine life and the fishing industry will also be severely affected in some places.

Climate change induced by increasing greenhouse gases is likely to affect crops differently from region to region. For example, average crop yield is expected to drop down to 50 per cent in Pakistan just as to the UKMO scenario whereas corn production in Europe is expected to grow up to 25 per cent in optimum hydrologic conditions.

More favourable effects on yield tend to depend to a large extent on realization of the potentially beneficial effects of carbon dioxide on crop growth and increase of efficiency in water use. Decrease in potential yields is likely to be caused by shortening of the growing period, decrease in water availability and poor vernalization.

In the long run, the climatic change could affect agriculture in several ways:

- Productivity, in terms of quantity and quality of crops
- Agricultural practices, through changes of water use and agricultural inputs such as herbicides, insecticides and fertilizers
- Environmental effects, in particular in relation of frequency and intensity of soil drainage, soil erosion, reduction of crop diversity
- Rural space, through the loss and gain of cultivated lands, land speculation, land renunciation, and hydraulic amenities.
- Adaptation, organisms may become more or less competitive, as well as humans may develop urgency to develop more competitive organisms, such as flood resistant or salt resistant varieties of rice.

They are large uncertainties to uncover, particularly because there is lack of information on many specific local regions, and include the uncertainties on magnitude of climate change, the effects of technological changes on productivity, global food demands, and the numerous possibilities of adaptation.

Most agronomists believe that agricultural production will be mostly affected by the severity and pace of climate change, not so much by gradual trends in climate. If change is gradual, there may be enough time for biota adjustment. Rapid climate change, however, could harm agriculture in many countries, especially those that are already suffering from rather poor soil and climate conditions, because there is less time for optimum natural selection and adaption.

PHYSICAL EVIDENCE FOR CLIMATIC CHANGE

Evidence for climatic change is taken from a variety of sources that can be used to reconstruct past climates. Reasonably complete global records of surface temperature are available beginning from the mid-late 19th century. For earlier periods, most of the evidence is indirect—climatic changes are inferred from changes in proxies, indicators that reflect climate, such as vegetation, ice cores, dendrochronology, sea level change, and glacial geology.

Historical and Archaeological Evidence

Climate change in the recent past may be detected by corresponding changes in settlement and agricultural patterns. Archaeological evidence, oral history and historical documents can offer insights into past changes in the climate. Climate change effects have been linked to the collapse of various civilizations.

Glaciers

Glaciers are considered among the most sensitive indicators of climate change, advancing when climate cools and retreating when climate warms. Glaciers grow and shrink, both contributing to natural variability and amplifying externally forced changes. A world glacier inventory has been compiled since the 1970s, initially based mainly on aerial photographs and maps but now relying more on satellites. This compilation tracks more than 100,000 glaciers covering a total area of approximately 240,000 km^2, and preliminary estimates indicate that the remaining ice cover is around 445,000 km^2. The World Glacier Monitoring Service collects data annually on glacier retreat and glacier mass balance from this data, glaciers worldwide have been found to be shrinking significantly, with strong glacier retreats in the 1940s, stable or growing conditions during the 1920s and 1970s, and again retreating from the mid 1980s to present.

The most significant climate processes since the middle to late Pliocene (approximately 3 million years ago) are the glacial and interglacial cycles. The present interglacial period (the Holocene) has lasted about 11,700 years. Shaped by orbital variations, responses such as the rise and fall of continental ice sheets and significant sea-level changes helped create the climate. Other changes, including Heinrich events, Dansgaard–Oeschger events and the Younger Dryas, however, show how glacial variations may also influence climate without the orbital forcing.

Glaciers leave behind moraines that contain a wealth of material—including organic matter, quartz, and potassium that may be dated—

recording the periods in which a glacier advanced and retreated. Similarly, by tephrochronological techniques, the lack of glacier cover can be identified by the presence of soil or volcanic tephra horizons whose date of deposit may also be ascertained.

Vegetation

A change in the type, distribution and coverage of vegetation may occur given a change in the climate; this much is obvious. In any given scenario, a mild change in climate may result in increased precipitation and warmth, resulting in improved plant growth and the subsequent sequestration of airborne CO_2. Larger, faster or more radical changes, however, may well result in vegetation stress, rapid plant loss and desertification in certain circumstances.

Ice Cores

Analysis of ice in a core drilled from a ice sheet such as the Antarctic ice sheet, can be used to show a link between temperature and global sea level variations. The air trapped in bubbles in the ice can also reveal the CO_2 variations of the atmosphere from the distant past, well before modern environmental influences. The study of these ice cores has been a significant indicator of the changes in CO_2 over many millennia, and continues to provide valuable information about the differences between ancient and modern atmospheric conditions.

Dendroclimatology

Dendroclimatology is the analysis of tree ring growth patterns to determine past climate variations. Wide and thick rings indicate a fertile, well- watered growing period, whilst thin, narrow rings indicate a time of lower rainfall and less-than-ideal growing conditions.

Insects

Remains of beetles are common in freshwater and land sediments. Different species of beetles tend to be found under different climatic conditions. Given the extensive lineage of beetles whose genetic makeup has not altered significantly over the millennia, knowledge of the present climatic range of the different species, and the age of the sediments in which remains are found, past climatic conditions may be inferred.

Sea Level Change

Global sea level change for much of the last century has generally been estimated using tide gauge measurements collated over long periods of time to give a long-term average. More recently, altimetre measurements — in combination with accurately determined satellite orbits — have provided an improved measurement of global sea level change. To measure sea levels prior to instrumental measurements, scientists have dated coral reefs that grow near the surface of the ocean, coastal sediments, marine terraces, ooids in limestones, and nearshore archaeological remains. The predominant dating methods used are uranium series and radiocarbon, with cosmogenic radionuclides being sometimes used to date terraces that have experienced relative sea level fall.

MEASUREMENT OF CLIMATE ELEMENTS

Measurement of temperature

Many surface air temperature records extend back to the middle part of the last century. The measurement of the surface air temperature is essentially the same now as it was then, using a mercury-in-glass thermometre, which can be calibrated accurately and used down to -39°C, the freezing point of mercury. For lower temperatures, mercury is usually substituted by alcohol. Maximum and minimum temperatures measured during specified time periods, usually 24 hours, provide useful information for the construction and analysis of temperature time series.

Analysis involves the calculation of averages and variances of the data and the identification, using various statistical techniques, of periodic variations, persistence and trends in the time series. Observations of temperature from surface oceans are also collected in order to construct time series. In recent decades, much effort has also been directed towards the measurement of temperature at different levels in the atmosphere. There are now two methods of measuring temperatures at different altitudes: the conventional radiosonde network; and the microwave-sounding unit (MSU) on the TIROS-N series of satellites. The conventional network extends back to 1958 and the MSU data to 1979.

Temperature is a valuable climate element in climate observation because it directly provides a measure of the energy of the system under inspection. For example, a global average temperature † reveals information about the energy content of the Earth-atmosphere system. A higher temperature would indicate a larger energy content. Changes in temperature indicate changes in the energy balance, the causes of which were discussed. Variations in temperature are also subject to less variability than other elements such as rainfall and wind. In addition, statistical analysis of temperature time series is often less complex than that associated with other series. Perhaps most importantly of all, our own perception of the state of the climate are intimately linked to temperature.

Measurement of Rainfall

Rainfall is measured most simply by noting periodically how much has been collected in an exposed vessel since the time of the last observation.

Care must be taken to avoid underestimating rainfall due to evaporation of the collected water and the effects of wind. Time series can be constructed and analysis performed in a similar manner to those of temperature.

The measurement of global rainfall offers an indirect or qualitative assessment of the energy of the Earth-atmosphere system. Increased heat storage will increase the rate of evaporation from the oceans (due to higher surface temperatures). In turn, the enhanced levels of water vapour in the atmosphere will intensify global precipitation. Rainfall is, however, subject to significant temporal and spatial variability, and the occurrence of extremes, and consequently, analysis of time series is more complex.

Measurement of Humidity

The amount of water vapour in the air can be described in at least 5 ways, in terms of:

- The water-vapour pressure;
- The relative humidity;
- The absolute humidity
- The mixing ratio
- The dewpoint.

A full account of these definitions may be found in Linacre. The standard instrument for measuring humidity is a psychrometre. This is a pair of identical vertical thermometres, one of which has the bulb kept wet by means of a muslin moistened by a wick dipped in water. Evaporation from the wetted bulb lowers its temperature below the air temperature (measured by the dry bulb thermometre).

The difference between the two measured values is used to calculate the air's water-vapour pressure, from which the other indices of humidity can be determined.

Measurement of Wind

Wind is usually measured by a cup anemometre which rotates about a vertical axis perpendicular to the direction of the wind. The exposure of wind instruments is important any obstruction close by will affect measurements. Wind direction is also measured by means of a vane, accurately balanced about a truly vertical axis, so that it does not settle in any particular direction during calm conditions.

Homogeneity

Non-climatic influences - inhomogeneities - can and do affect climatic observations. Any analyst using instrumental climate data must first assess the quality of the observations. A numerical series representing the variations of a climatological element is called homogeneous if the variations are caused only by fluctuations in weather and climate.

Leaving aside the misrecording of data, the most important causes of inhomogeneity are:

- Changes in instrument, exposure and measuring technique (for example, when more technologically advanced equipment is introduced);
- Changes in station location (i.e. when equipment is moved to a new site);
- Changes in observation times and methods used to calculate daily averages; and
- Changes in the station environment, particularly urbanisation (for example, the growth of a city around a pre-existing meteorological station).

When assessing the homogeneity of a climate record, there are three major sources of information: the variations evident in the record itself; the station history; and nearby station data. Visual examination and statistical analysis of the station record may reveal evidence of systematic changes or unusual behaviour which suggest inhomogeneity. For example, there may be a step-change in the mean, indicating a change in station location. A steady trend may indicate a progressive change in the station environment, such as urbanisation. An extreme value may be due to a typing error.

Often these inhomogeneities may be difficult to detect and other evidence is needed to confirm their presence. One source of evidence is the station history, referred to as metadata. The station history should include details of any changes of location of the station, changes in instrumentation or changes in the timing and nature of observation. Very often, though, actual correction factors to observation data containing known inhomogeneities will be difficult to calculate, and in these cases, the record may have to be rejected.

The third approach to homogenisation involves empirical comparisons between stations close to each other. Over time scales of interest in climate change studies, nearby stations (i.e. within 10km of each other) should be subject to similar changes in monthly, seasonal and annual climate.

The only differences should be random. Any sign of systematic behaviour in the differences (e.g. a trend or step-change) would suggest the presence of inhomogeneities. In light of the foregoing discussion on homogeneity, careful attention has to be paid to eliminating sources of non-climatic error when constructing large-scale record, such as the global surface air temperature time series. This, and similar records including sea surface temperatures, rely on the collection of millions of individual observations from a huge network made up of thousands of climate stations.

A number of these have been reviewed by Joneset al.. The effects of urbanisation (the artificial warming associated with the growth of towns and cities around monitoring sites) were considered to be the greatest source of inhomogeneity, but even this, it was concluded, accounts for at most a 0.05°C warming (or 10 per cent of the observed warming) over the last 100 years. Conrad and Pollak (1962), Folland et al. (1990), and Jones et al. provide useful references investigating the problems of homogeneity and data reliability of instrumental records of climate data.

Statistical Analysis of Instrumental Records

Once climate data has been collected and corrected for inhomogeneities, it will need to be analysed. The aim of any statistical analysis is to identify systematic behaviour in a data set and hence improve understanding of the processes at work to compliment the theory. Statistical analysis is a search for a signal in the data that can be distinguished from the background noise In climate change research that signal will be a periodic variation, a quasi-periodic variation, a trend, persistence or extreme events in the climate element under analysis.

Before undertaking a statistical analysis of a climate record, a number of questions about the task in hand should be asked.

1. What is the purpose of the analysis?

In its simplest form, the statistical analysis may be:

- Descriptive; or
- Investigative.

Descriptive analyses set out solely to document particular aspects of the variations present in the data set (the signal). Indices calculated will include the mean and variance (or standard deviation). The occurrence of extreme events, cycles and trends will also be noted. Significance testing is crucial to this category of analysis. Significance testing establishes whether or not the variation under consideration is different from what one would expect to arise in a random time series.

Investigative analyses set out to test a pre-defined hypothesis. The hypothesis should a priori have a sound physical basis. "Does the time series contain an El Niño cycle?" would be an example of a hypothesis that could be investigated.

2. What is the most appropriate data set to use?

Any data set used for a statistical analysis should be:

- Representative of the relevant physical processes;
- Sufficient in quantity to support the statistical method(s) to be used; and
- Accurate and reliable (homogeneous).

To investigate the impact of El Niño † on drought in eastern Australia, it is necessary to firstly identify a representative indicator of El Niño, such as sea surface temperatures in the SE Pacific. Secondly, a reliable indicator of drought in eastern Australia is required, for example, rainfall. The data set would need to be of sufficient duration to permit the testing for a relationship on the time scale under consideration, i.e. does El Niño cause drought in eastern Australia? Since El Niño has a quasi-periodicity of 2 to 5 years, then a data set of length at least 7 to 10 times this duration (i.e. up to 50 years) is required to have confidence in the statistical methods. To investigate longer-term trends, the data requirement becomes more stringent.

3. What is the most appropriate technique to use and how should it be applied?

Often it will be clear as to which statistical method of analysis is required. However, its application may be less straightforward. The nature of the data may determine whether or not a particular technique is valid (or, at least the way in which the technique is applied). For example, if the data are not normally distributed † then this may invalidate assumptions on which the technique is based. What ever technique is used, it goes without saying that testing statistical significance must be a critical concern.

Barry and Perry offer a detailed introduction of the mathematical aspects of statistical analysis, with many useful examples. Other useful references are provided by Gani and Godske. Before concluding, however, a couple of points need illustrating. Firstly, much of what has been said about statistical analysis of instrumental records applies equally well to the study of palaeoclimatology, and to the reconstruction of past climates from proxy data. Secondly, the statistical analysis of climate data serves to compliment and support theories developed to explain the causes (and effects) of climate change. Statistical associations do not prove cause and effect for they are solely based upon the laws of probability. When analysing and interpreting climate data in the effort to aid understanding of the causes of climate change, it is necessary to bare this in mind.

Palaeoclimate Reconstruction from Proxy Data

Climate varies over different time scales, from years to hundreds of millions of years, and each periodicity is a manifestation of separate forcing mechanisms. In addition, different components of the climate system change and respond to forcing factors at different rates; in order to understand the role such components play in the evolution of climate it is necessary to have a record considerably longer than the time it takes for them to undergo significant changes.

Palaeoclimatology is the study of climate and climate change prior to the period of instrumental measurements. A geological chronology of Earth history is provided in the appendix, which will provide a useful time frame for the material discussed both in the remainder.

Instrumental records span only a tiny fraction (<10-7) of the Earth's climatic history and so provide a inadequate perspective on climatic variation and the evolution of the climate today and in the future. A longer perspective on climate variability can be obtained by the study of natural phenomena which are climate-dependent. Such phenomena provide a proxy record of the climate.

Many natural systems are dependent on climate, and from these it may be possible to derive palaeoclimatic information from them. By definition, such proxy records of climate all contain a climatic signal, but that signal may be weak and embedded in a great deal of random (climatic) background noise. In essence, the proxy material has acted as a filter, transforming climate conditions in the past into a relatively permanent record. Deciphering that record is often a complex business.

Principle sources of proxy data for palaeoclimatic reconstructions:

- Glaciological (Ice Cores)
  - Oxygen isotopes
  - Physical properties
  - Trace element and microparticle concentrations

Geological

A. Sediments:

1. Marine (ocean sediment cores)
   i) Organic sediments (planktonic and benthic fossils)
      Oxygen isotopes
      Faunal and floral abundances
      Morphological variations
   ii) Inorganic sediments
      Mineralogical composition and surface texture
      Distribution of terrigenous material
      Ice-rafted debris
      Geochemistry
2. Terrestrial
   i) Periglacial features
   ii) Glacial deposits and erosional features
   iii) Glacio-eustatic features (shorelines)
   iv) Aeolian deposits (sand dunes)
   v) Lacustrine deposits/varves (lakes)

B. Sedimentary Rocks:

- Facies analysis
- Fossil/microfossil analysis
- Mineral analysis
- Isotope geochemistry

C. Biological:

- Tree rings (width, density, isotope analysis)
- Pollen (species, abundances)
- Insects

D. Historical:

- Meteorological records
- Parameteorological records (environmental indicators)
- Phenological records (biological indicators)

The major types of proxy climatic data available are listed. Each proxy material differs according to: a) its spatial coverage; b) the period to which it pertains; and c) its ability to resolve events accurately in time. Some proxy records, for example ocean floor sediments, reveal information about long periods of climatic change and evolution (107 years), with a low-frequency resolution (103 years). Others, such as tree rings are useful only during the last 10,000 years at most, but offer a high frequency (annual) resolution. The choice of proxy record (as with the choice of instrumental record) very much depends on what physical mechanism is under review. As noted, climate responds to different forcing mechanisms over different time scales, and proxy materials will contain necessary climatic information on these to a greater or lesser extent, depending on the three factors mentioned.

Other factors that have to be considered when using proxy records to reconstruct palaeoclimates include the continuity of the record and the accuracy to which it can be dated. Ocean sediments may provide continuous records for over 1 million years (Ma) but typically they are hard to date using existing techniques. Usually there is an uncertainty of +/– 5 per cent of the record's true age. Ice cores are easier to date but may miss layers from certain periods due to melting and wind erosion. Glacial deposits are highly episodic in nature, providing evidence only of discrete events in the past. Different proxy systems also have different levels of inertia with respect to climate, such that some systems may vary exactly in phase with climate whereas others lag behind by as much as several centuries.

Like climate construction from instrumental records, palaeoclimate reconstruction may be considered to proceed through a number of stages. The first stage is that of proxy data collection, followed by initial analysis and measurement. This results in primary data. The next stage involves the calibration of the data with modern climate records. In this, the uniformitarian principle is assumed, whereby contemporary climatic variations form a modern analogue for palaeoclimatic changes. It is important to be aware, however, of the possibility that palaeo-environmental conditions may not have modern analogues. The calibration may be only qualitative, involving subjective assessment, or it may be highly quantitative. The secondary data provide a record of past climatic variation. The third stage is the statistical analysis of this secondary data. The palaeoclimatic record is now statistically described and interpreted, providing a set of tertiary data.

Obviously not exhaustative. Bradley offers an excellent review of the various proxy methods and techniques employed to reconstruct Quaternary † palaeoclimatic change, whilst Frakes provides a useful commentary on the evidence for pre-Quaternary climates spanning most of geologic time. In the following, some of the more widely used proxy techniques will be reviewed. In all of the accounts, attention should be paid to the issues of reliability, dating, interpretation and meaning for all forms of climate reconstruction.

Historical Records

Historical records have been used to reconstruct climates dating back several thousand years (i.e. for most of the Holocene). Historical proxy data can be grouped into three major categories. First, there are observations of weather phenomena per se, for example the frequency and timing of frosts or the occurrence of snowfall. Secondly, there are records of weather-dependent natural or environmental phenomena, termed parameteorological phenomena, such as droughts and floods. Finally, there are phenological records of weather- dependent biological phenomena, such as the flowering of trees, or the migration of birds.

Major sources of historical palaeoclimate information include: ancient inscriptions; annals and chronicles; government records; estate records; maritime and commercial records; diaries and correspondence; scientific or quasi-scientific writings; and fragmented early instrumental records.

There are a number of major difficulties in using this kind of information. First, it is necessary to determine exactly what the author meant in describing the particular event. How severe was the "severe" frost? What precisely does the term drought refer to? Content analysis - a standard historical technique - has been used to assess, in quantitative terms, the meaning of key climatological phrases in historical accounts. This approach involves assessment of the frequency of use made of certain words or phrases by a particular author. Nevertheless, the subjectivity of any personal account has to be carefully considered. Very often, the record was not kept for the benefit of the future reader, but to serve some independent purpose. During much of the dynastic era in China, for example, records of droughts and floods would be kept in order to gain tax exemptions at times of climatic adversity.

Secondly, the reliability of the account has to be assessed. It is necessary to determine whether or not the author had first-hand evidence of the meteorological events. Thirdly, it is necessary to date and interpret the information accurately. The representativeness of the account has to be assessed. Was the event a localised occurrence or can its spatial extent be defined by reference to other sources of information? What was the duration of the event? a day? a month? a year?

Finally, the data must, as with all proxy records, be calibrated against recent observations and cross-referenced with instrumental data This might be achieved by a construction of indices (e.g. the number of reports of frost per winter) which can be statistically related to analogous information derived from instrumental records.

Ice Cores

As snow and ice accumulates on polar and alpine ice caps and sheets, it lays down a record of the environmental conditions at the time of its formation. Information concerning these conditions can be extracted from ice and snow that has survived the summer melt by physical and chemical means. When melting does occur, the refreezing of meltwater can provide a measure of the summer conditions.

Palaeoclimate information has been obtained from ice cores by three main approaches.

These involve the analysis of:

- Stable isotopes of water;
- Dissolved and particulate matter in the firn † and ice; and
- The physical characteristics of the firn and ice, and of air bubbles trapped in the ice. Each approach has also provided a means of dating the ice at particular depths in the ice core.

Stable Isotope Analysis

The basis for palaeoclimatic interpretations of variations in the stable isotope † content of water molecules is that the vapour pressure of $H2^{16}O$ is higher than that of $H2^{18}O$. Evaporation from a water body thus results in a vapour which is poorer in ^{18}O than the initial water; conversely, the remaining water is enriched in ^{18}O. During condensation, the lower vapour pressure of the $H2^{18}O$ ensures that it passes more readily into the liquid state than water vapour made up of the lighter oxygen isotope. During the poleward transportation of water vapour, such isotope fractionation continues this preferential removal of the heavier isotope, leaving the water vapour increasingly depleted in $H2^{18}O$. Because condensation is the result of cooling, the greater the fall in temperature, the lower the heavy isotope concentration will be. Isotope concentration in the condensate can thus be considered as a function of the temperature at which condensation occurs. Water from polar snow will thus be found to be most depleted in $H2^{18}O$.

This temperature dependency allows the oxygen isotope content of a ice core to provide a proxy climate record. The relative proportions of ^{16}O

and ^{18}O in an ice core are expressed in terms of departures, $\delta^{18}O$, from the Standard Mean Ocean Water (SMOW) standard, such that:

$$\frac{\delta\, O = (O/O)_{sample} - (O/O)_{SMOW} \times 10\%}{\left(^{18}O/^{16}O\right)_{SMOW}}$$

All measurements are made using a mass spectrometre and results are normally accurate to within 0.1‰ (parts per mille). A $\delta^{18}O$ value of –10‰ indicates a sample with an $^{18}O/^{16}O$ ratio 1 per cent or 10‰ less than SMOW. For most palaeoclimate reconstructions, typical values for $\delta^{18}O$ obtained from ice cores range between -10 and -60‰.

Similar palaeoclimatic studies can be carried out using isotopes of hydrogen (^{1}H and ^{2}H (Deuterium)), but these are rarer in nature and the laboratory techniques involved are more complex.

Physical and Chemical Characteristics of Ice Cores

The occurrence of melt features in the upper layers of ice cores are of particular palaeoclimatic significance. Such features include horizontal ice lenses and vertical ice glands which have resulted from the refreezing of percolating water. They can be identified by their deficiency in air bubbles.

The relative frequency of melt phenomena may be interpreted as an index of maximum summer temperatures or of summer warmth in general. Other physical features of ices cores which offer information to the palaeoclimatologist include variations in crystal size, air bubble fabric and crystallographic axis orientation.

Finally, variations of particulate matter, particularly calcium, aluminium, silicon and certain atmospheric aerosols can also be used as proxy palaeoclimatic indicators.

Dating Ice Cores

One of the biggest problems in any ice core study is determining the age- depth relationship. Many different approaches have been used and it is now clear that fairly accurate time scales can be developed for the last 10,000 years. Prior to that, there is increasing uncertainty about ice age. The problem lies with the fact that the age-depth is highly exponential, and ice flow models are often needed to determine the ages of the deepest sections of ice cores. For example, the upper 1000m of a core may represent 50,000 years, whilst the next 50m may span another 100,000 year time period, due to the severe compaction, deformation and flow of the ice sheet in question.

Radio isotope dating †, using ^{210}Pb (lead), ^{32}Si (silicon), ^{39}Ar (argon) and ^{14}C (carbon) have all been used with varying degrees of success, over different time scales, to determine the age of ice cores.

Certain components of ice cores may reveal quite distinct seasonal variations which enable annual layers to be identified, providing accurate time scales for the last few thousand years. Such seasonal variations may be found in $\delta^{18}O$ values, trace elements and microparticles (Hammer *et al.*, 1978).

Where characteristic layers of known ages can be detected, these provide valuable chronostratigraphic markers against which other dating methods can be verified. So-called reference horizons have resulted from major explosive volcanic eruptions. These inject large quantities of dust and gases (principally sulphur dioxide) into the atmosphere, where they are globally dispersed. The gases are converted into aerosols (principally of sulphuric acid) before being washed out in precipitation. Hence, after major eruptions, the acidity of snowfall increases significantly above background levels. By identifying highly acidic layers (using electrical conductivity) resulting from eruptions of known age, an excellent means of checking seasonally based chronologies is available.

Dendroclimatology

The study of the annual growth of trees and the consequent assembling of long, continuous chronologies for use in dating wood is called dendrochronology. The study of relationships between annual tree growth and climate is called dendroclimatology. Dendroclimatology offers a high resolution (annual) form of palaeoclimate reconstruction for most of the Holocene.

The annual growth of a tree is the net result of many complex and interrelated biochemical processes. Trees interact directly with the microenvironment of the leaf and the root surfaces. The fact that there exists a relationship between these extremely localised conditions and larger scale climatic parametres offers the potential for extracting some measure of the overall influence of climate on growth from year to year. Growth may be affected by many aspects of the microclimate: sunshine, precipitation, temperature, wind speed and humidity. Besides these, there are other non- climatic factors that may exert an influence, such as competition, defoliators and soil nutrient characteristics.

There are several subfields of dendroclimatology associated with the processing and interpretation of different tree-growth variables. Such variables include tree-ring width (the most commonly exploited information source, e.g. Briffa and Schweingruber,), densitometric parametres and chemical or isotopic variables.

A cross section of most temperate forest tree trunks † will reveal an alternation of lighter and darker bands, each of which is usually continuous around the tree circumference. These are seasonal growth increments produced by meristematic tissues in the cambium of the tree. Each seasonal increment consists of a couplet of earlywood (a light growth band from the early part of the growing season) and denser latewood (a dark band produced towards the end of the growing season), and collectively they make up the tree ring. The mean width of the tree ring is a function of many variables, including the tree species, tree age, soil nutrient availability, and a whole host of climatic factors.

The problem facing the dendroclimatologist is to extract whatever climatic signal † is available in the tree-ring data from the remaining background "noise".

Whenever tree growth is limited directly or indirectly by some climate variable, and that limitation can be quantified and dated, dendroclimatology can be used to reconstruct some information about past environmental conditions.

Only for trees growing near the extremities of their ecological amplitude † , where they may be subject to considerable climatic stresses, is it likely that climate will be a limiting factor. Commonly two types of climatic stress are recognised, moisture stress and temperature stress.

Trees growing in semi-arid regions are frequently limited by the availability of water, and dendroclimatic indicators primarily reflect this variable. Trees growing near the latitudinal or altitudinal treeline are mainly under growth limitations imposed by temperature; hence dendroclimatic indicators in such trees contain strong temperature signals.

Furthermore, climatic conditions prior to the growth period may precondition physiological processes within the tree and hence strongly influence subsequent growth. Consequently, strong serial correlation or autocorrelation may establish itself in the tree-ring record. A specific tree ring will contain information not just about the climate conditions of the growth year but information about the months and years preceding it.

Several assumptions underlie the production of quantitative dendroclimatic reconstructions. First, the physical and biological processes which link toady's environment with today's variations in tree growth must have been in operation in the past. This is the principle of uniformitarianism.

Second, the climate conditions which produce anomalies in tree-growth patterns in the past must have their analogue during the calibration period.

Third, climate is continuous over areas adjacent to the domain of the tree- ring network, enabling the development of a statistical transfer function relating growth in the network to climate variability inside and outside of it. Finally, it is assumed that the systematic relationship between climate as a limiting factor and the biological response can be approximated as a linear mathematical expression. Fritts provides a more exhaustive review of the assumptions involved in the use of dendroclimatology.

The general approach taken in dendroclimatic reconstruction is:

- To collect (sample) data from a set of trees (within a tree population) which have been selected on the basis that climate (e.g. temperature, moisture) should be a limiting factor;
- To assemble the data into a composite site chronology by cross-dating the individual series after the removal of age effects by standardization †. This master chronology increases the (climate) signal to (non-climate) noise ratio;
- To build up a network of site chronologies for a region;
- To identify statistical relationships between the chronology times series and the instrumental climate data for the recent period - the calibration period;
- To use these relationships to reconstruct climatic information from the earlier period covered by the tree-ring data, and;
- Finally, to test, or verify, the resulting reconstruction against independent data.

Bradley gives a full account of the methods (1 to 6 above) of palaeoclimate reconstruction from tree-ring analysis. This approach may be applied to all the climate-dependent tree-growth variables, specifically tree-ring width, but also wood density and isotopic measurements.

The latewood of a tree ring is much denser than the earlywood and interannual variations contain a strong climatic signal. Density variations are particularly valuable in dendroclimatology because they to not change significantly with tree age, and the process of standardisation (removal of growth function) can therefore be avoided.

The use of isotopic measurements in dendroclimatology also avoids the need for a standardisation process. The basic premise of isotope dendroclimatology is that since $^{18}O/^{16}O$ and D/H (deuterium/hydrogen) variations in meteoric (atmospheric) waters are a function of temperature, tree growth which records such isotope variations should preserve a record of past temperature fluctuations.

Unfortunately, isotope fractionation effects within the tree, which are themselves temperature dependent, will create problems associated with this technique.

Ocean Sediments

Billions of tonnes of sediment accumulate in the ocean basins every year, and this may be indicative of climate conditions near the ocean surface or on the adjacent continents. Sediments are composed of both biogenic (organic) and terrigenous (inorganic) materials.

The biogenic component includes the remnants of planktonic (surface ocean-dwelling) and benthic (deep-water- or sea floor-dwelling) organisms which provide a record of past climate and oceanic circulation. Such records may reveal information about past surface water temperatures, salinity, dissolved oxygen and nutrient availability. By contrast, the nature and abundance of terrigenous materials provides information about continental humidity-aridity variations, and the intensities and directions of winds. Ocean sediment records have been used to reconstruct palaeoclimate changes over a range of time scales, from thousands of years to millions and even tens of millions of years in the past.

CHANGES IN GLACIERS, ICE SHEETS AND ICE SHELVES

Glaciers store water over relatively long timescales compared to rivers and lakes—hundreds to a few thousands of years. Ice sheets store water for even longer—tens of thousands of years. But the shorter glacier timescales are comparable to human timescales, so people notice how glaciers change, and these changes have obvious impacts on the human environment. Many river systems depend on glacier melt, which maintains the water supply through the summer.

As glaciers shrink, so does the frozen water supply they store. This is one of the reasons why it is important to measure how and understand why glaciers change over time.

Glacier melt water eventually makes its way to the ocean, where it can affect global sea level. As terrestrial ice masses grow, sea level falls; and as masses shrink, sea level rises. At the last glacial maximum, about 18,000 years ago, the growth of ice sheets and glaciers caused sea level to lower by about 120 metres (395 feet). Most of that change was due to the formation of large ice sheets in northern North America and Europe, but mountain glaciers, too, had their role.

Glacier change may be quantified by a change in length, by a change in the surface area covered, or most rigourously by the glacier's mass balance, the difference between the volume of ice accumulated from snowfall and the volume of ice lost to melting or iceberg calving in a year. A glacier is "in balance" when gains and losses are equal. Because glaciers and ice sheets are remote and are difficult to observe in detail, the errors involved in estimating how they change are large. The best estimates indicate that while the Greenland and Antarctic ice sheets are near balance, glaciers and ice caps are shrinking worldwide, contributing as much as 0.25 millimetres (0.01 inch) per year to global sea-level rise.

Modern Glacier Retreat

Mountain glaciers are always changing because they tend towards balance with ever-changing snowfall and melting rates. Some special glaciers change volume for other reasons . "Surge-type glaciers" experience rapid speed-ups that move ice downstream quickly, thinning the glacier, followed by quiescent periods in which the ice thickens once again. "Tidewater glaciers" end as a floating ice tongue in a lake or bay, where interaction with the lake bottom or seafloor, and iceberg calving events may complicate their flow.

Yet despite all the possible complications, glaciologists conclude that worldwide, glaciers are retreating rapidly as climate warms. For example, the glaciers of Mount Kilimanjaro, about which Hemingway wrote in 1938, are predicted to be gone by 2020. Ice first started to accumulate on that mountain nearly 12,000 years ago. Alaskan glaciers, measured with laser from a low- flying airplane, are thinning rapidly and contributing as much as 0.14 millimetres (0.005 inches) per year to sea level. The thinning rate has increased within the last decade. In Antarctica, ice-shelf collapse along its northward- reaching peninsula has also been linked to ongoing warming. Glacier retreat also is being monitored in the Himalayas, where increased melting can lead to flood risks for people living downstream. Around the world, glaciers are retreating in response to warming weather, and water resources and sea level will continue to be affected as they do.

Antarctic Ice Shelf Breakups

Iceberg calving is an efficient and normal process by which floating ice shelves maintain a steady mass balance over many years; but it also can be an important indicator of climate change. While temperatures in the interior of Antarctica have remained fairly steady since scientists arrived on the continent, the Antarctic Peninsula has warmed about 2.5°C over the last half of the twentieth century.

An ice shelf is a floating ice mass that is attached to the coast along at least one edge. The year 2002 brought notable events involving Antarctic ice shelves. Two massive icebergs broke off from the Ross Ice Shelf, a large sheet of glacial ice and snow extending from the Antarctic mainland into the southern Ross Sea. In a different event, an enormous piece broke away from the Larsen B Ice Shelf, located on the eastern side of the Antarctic Peninsula, in the largest single event in a 30-year series of ice-shelf retreats.

In the Larsen B Ice Shelf event, which reduced the shelf to a size not seen for about 12,000 years, a 3,275-square-kilometre (1,260-square-mile) sector disintegrated over the course of 35 days. The rapid break-up of the floating ice mass, which had survived thousands of years of climate variations, came at the end of one of the warmest summers on record.

When summertime temperatures rose close to and above freezing, snow on the surface of the Larsen Ice Shelf began to melt, eventually forming kilometres-wide ponds. That water is believed to have filled crevasses, which are deep, wedge-shaped cracks in the surface of the ice. Because water is denser than ice, it exerts extra pressure on the sides of a crevasse.

Four Moderate Resolution Imaging Spectroradiometre (MODIS) satellite images show the progressive breaking apart of a 3,275-square-kilometre (1,260- square-mile) northern section of the Larsen B Ice Shelf in Antarctica. The top left image was taken January 31, 2002, and the bottom right image was taken March 5, 2002. The snowy ice-shelf surface is clearly visible to the left of each frame, with the dark ocean waters to the right.

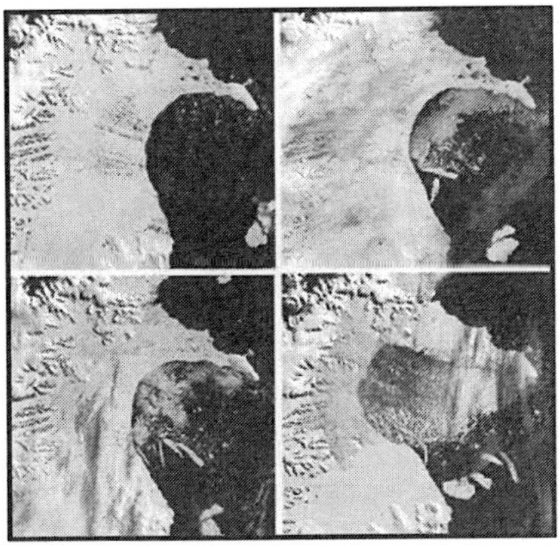

The mountains of the Antarctic Peninsula curve along the western (left) side of each frame. Meltwater ponds form patches and stripes on the surface of the shelf. A few small icebergs are floating in the dark ocean surface in pre-breakup images (top); the icebergs in the far lower right of the frames had calved 2 years earlier from an ice shelf farther south. After March 2002, long narrow icebergs and fragments too small to be resolved by the satellite were nearly all that remained of the Larsen B Ice Shelf.

Cause it to crack down to the bottom of the shelf. When the meltwater-enhanced fractures became widespread, they weakened the ice shelf, allowing it to quickly break apart.

Scientists cannot yet attribute specific ice events to global warming, due to lack of specific long-term data that would yield definitive causal relationships. Recent events must be viewed in the context of long-term, natural processes as well as the relatively short-term body of data collected since the Industrial Revolution.

Records of Climate Change

In addition to being affected by climate change, glaciers and ice sheets are archives of climate-change data. Each winter, new snow falls on the surface of the glacier. Whatever snow does not melt during the following summer will be buried by more snow the next winter. This "old snow" is called firn. The frozen water molecules and air trapped in the firn record the chemistry and temperature of the water vapour from which the snow formed and the atmosphere from which it fell.

As each year's firn layer is buried, that climate record is buried as well. The firn layers move down from the surface and are compressed as new layers pile on top. Eventually, the firn becomes dense, glacier ice. Most of the air has been squeezed out of the ice, but a few bubbles remain. When glaciologists drill down through the ice, they are drilling backward in time; consequently, ice cores drilled from glaciers and ice sheets reveal both regional and global climate trends.

The longest ice-core climate record comes from Vostok, a Russian scientific station in East Antarctica. The ice sheet is 3,623 metres (2.2 miles) thick at Vostok Station, and reaches back in time more than 400,000 years. The ice and the bubbles trapped within it have been studied by glaciologists from Russia, France, and the United States.

The ice core reveals that global atmospheric carbon dioxide (CO_2), methane (CH_4), and dust content rise and fall as global temperature and ice volume change. When the climate is warm, atmospheric CO_2 and CH_4 concentrations are large, and when climate is cool, those gases are less abundant. Atmospheric dust concentration changes in an opposite sense, indicating that warm, interglacial atmospheres are relatively moist, whereas in glacial times, the global atmosphere is relatively dry. These records also show that the present-day CO_2 level is larger than it was in the past warm times between glaciations.

Ice cores retrieved from mountain glaciers record both global and local changes. For example, ice cores from glaciers in Peru and Bolivia (in South America) are being used to establish how tropical climate and the Amazonian rain forest change as Earth's climate warms and cools. These ice cores also contain long records of the El Niño Southern Oscillation. Tropical glaciers are melting fast as climate warms, and as that happens, their contribution to water resources decreases, their contribution to global sea level increases, and a valuable climate archive is lost. The first photographs made for World View of Global Warming, ten years ago, were of glaciers in Antarctica and Peru. This page shows a selection of the locations where we have documented glacier and ice cap retreat— a small set of images illustrating the overwhelming evidence from hundreds of glaciers and ice caps on every continent that global warming is severely affecting the water and glacial cycles of the planet. This is a profound change that unlike natural cycles like the Little Ice Age of the 16th and 17th centuries is proceeding very rapidly and appears tied to no natural cycles. The best correlation for this change to all but a handful of the 160,000 land glaciers and parts of the great ice caps is to rising atmospheric temperatures tied to increasing amounts of greenhouse gases.

The largest implication of this loss of glaciers is not the change in scenery, but the fact that the seasonal meltwater from glaciers, especially in Asia and South America, is the life support for billions of people. Large cities like Lima get much of their water from glaciers. In other parts of the world, glacier water keeps streams cool and full for salmon and other important wildlife. And as more and more water reaches the ocean, it is increasing sea level at a faster rate — which threatens every coastal city and shoreline. New scientific projections show at least a three foot (one metre) rise in ocean levels by the end of this century, part of which is also due to the expansion of warming sea water. This will inundate rice fields and estuaries that feed billions, and push into the heart of the worlds largest cities — and make each storm a threat of more devastating waves and surges.

The disintegrating face of the Müller Ice Shelf, Lallemand Fjord, Antarctic Peninsula, 67° South, April 2, 1999. This small shelf, fed by glaciers from the Loubet Coast, has been receding recently after growing over a 400-year cooling period. Like other receding ice shelves such as the larger Larsen, it may be a sensitive monitor of rising regional temperatures. The Larsen Ice Shelf lost a 1200 square mile section early in 2002, prompting some glaciologists to be concerned that even the giant Ross Ice Shelf could be at risk.

Fig. *Ice Under Fire: Antarctica*

This mile-long ice cliff of Marr Ice Piedmont, Anvers Island, has receded about 500 metres since the mid 1960s. The cliff's previous position was to the left of the line of ice floating in the harbour and extended to the headland at the extreme upper left. The regional temperature has increased 5° C in winter over the past 50 years. This reduces seasonal icepack, disrupts growth of krill and changes conditions on penguin rookeries.

Fig. *Ice Under Fire: Arctic*

Geophysicist Tom Osterkamp indicates ground level when he installed this temperature probe pipe near Denali Park, 15 years ago. Alaska permafrost temperature has increased 0.5° to 1.5°C since 1980, when Osterkamp began measuring ground temperature at 30 sites. Many of those sites show thermokarsting like this. Serious effects include forest damage, sinking roads and buildings, eroding tundra riverbanks, changes in tundra vegetation, and increased carbon dioxide and methane emissions from thawed peat.

Just how serious a threat thawing tundra can be is being measured in Northern Sweden. Rising atmospheric temperatures are causing the disappearance of permafrost and its replacement with marshland or open water. At Stordalen Mire near Abisko National Park, where careful records and aerial photos were available for comparison, the amount of wetland increased more than 50 per cent at the expense of higher dry land which had been underlain by permafrost.

Fig. *Ice Under Fire: Mountain Glaciers*

Study leader Patrick Crill reports this has increased the emission of methane from the mire, a powerful greenhouse gas, by as much as two-thirds over 1970 emissions. A 2004 study by Torben Christensen and others directly measured hydrocarbon emissions from the Mire, and found that almost all the emissions were methane. The scientists express concern that methane emissions seem to increase with temperature and could be a positive feedback creating even more atmospheric warming.

Photographer Gary Braasch holding a 1932 photo of Broggi glacier near Huascaran in the Peruvian Andes, while rephotographing this receding glacier in 1999. Glaciers everywhere in the world (with a very few exceptions) have been shrinking throughout the 20th Century, a prime signal of rapid global warming. Loss of tropical glaciers is particularly rapid. This glacier, previously photographed by the Austrian Hans Kinzl, receded about one kilometre in 67 years.

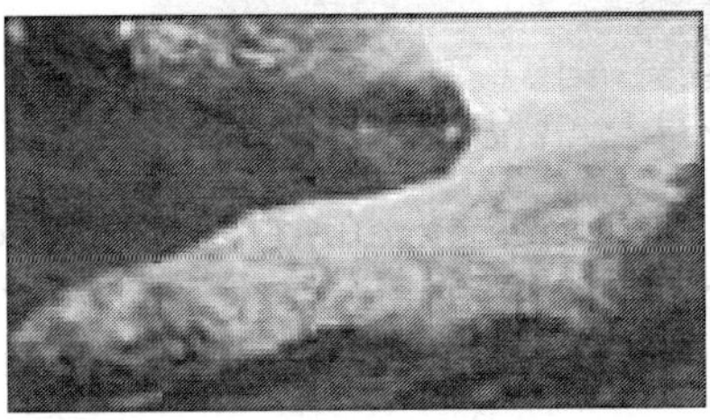

Glacier Ururashraju, at about 15,000 feet in Cordillera Blanca of Peru. Photographed in 1986 by Peruvian glaciologist Alcides Ames, whose studies and direction allowed Gary Braasch to rephotograph it in 1999. Retreat of about 500 M. Ames studies confirms the very rapid deglaciation of the Cordillera Blanca, which is the most ice-covered mountain range in the tropics.

Grinnell glacier, Glacier National Park, USA, which has receded almost out of sight. Photographed by a park photographer in 1911 and by Gary Braasch in 2000. Estimates by US National Park Service scientist Dan Fagre are that all but a few of the 30 glaciers in this Northern Rocky Mountains park will be gone by mid-century.

Perhaps the most dramatic glacier withdrawal has been in the Alps, where it has occured in full view of residents, tourists and scientists. This is a 1859 etching of the Rhone glacier in the Kanton of Valais, Switzerland, when the ice filled the valley right to the tiny crossroads of Gletsch. In 2001 the glacier was nearly out of sight, 2.5 km distant and 450 M higher.

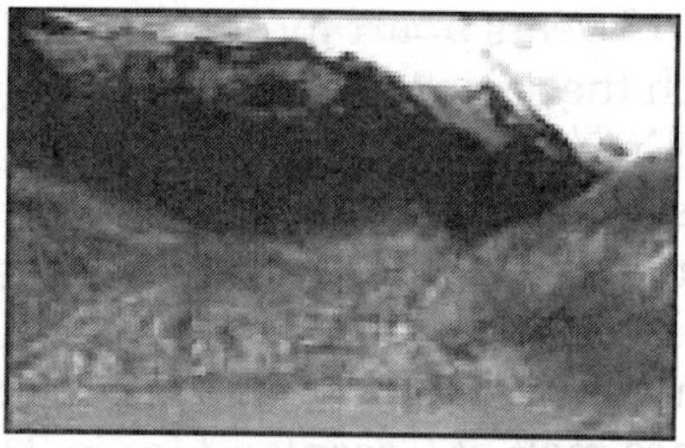

The Pasterze, Austria's longest glacier, was about 2 kilometres longer in the 19th C. but is now completely out of sight from this overlook on the Grossglockner High Road.

The Margaritzen-Strausee, a dammed artificial lake, now is in the place where the glacier terminus was in 1875. Measurements of the Pasterze began in 1889 and it has been pulling back the entire time, in approximate step with regional temperatures that have been increasing. The glacier is now about eight Km long and loses about 15 metres per year. However in 2003 the Pasterze decreased 30 metres in length and 6.5 metres in thickness. Alaska's glaciers are receding at twice the rate previously thought, according to a new study published in the July 19, 2002 Science journal.

These two images show Portage Glacier, near Anchorage, Alaska, in about 1950 and in July 2001. The ice has pulled back nearly out of sight.

Glaciers in the Northwest United States have also been shrinking. Studies by the Climate Impacts Group at University of Washington show regional temperature has been 1.5° F warmer in the 20th century, with rising snow lines, decreasing mountain snowpack, and earlier spring run-off.

These photos of Mt. Hood Oregon comparing late season snow and ice only 18 years apart indicate the problem: much less late summer ice from which the region gets water for irrigation, drinking, and fish habitat.

Greenland's huge icecap, second only to Antarctica, is also showing signs of change, although measurements are preliminary. Outflow glaciers like this one on the central east coast, as measured by NASA airborne radar and laser, appear to be thinning and flowing more rapidly. The National Climate Data Centre (NOAA) reports that 2002 saw the greatest measured surface melt of Greenland ice in 24 years of satellite records.

Two deep ice cores from there provide a detailed Northern Hemisphere climate record extending beyond the last ice age. Around the globe, sea level is more than 6 inches higher than it was 100 years ago, due primarily to warmer sea water and run-off from melting glaciers. The rate of rise is now 12 inches per century and increasing. The IPCC projected in its 2007 report a sea level up to 2 feet higher by 2100.

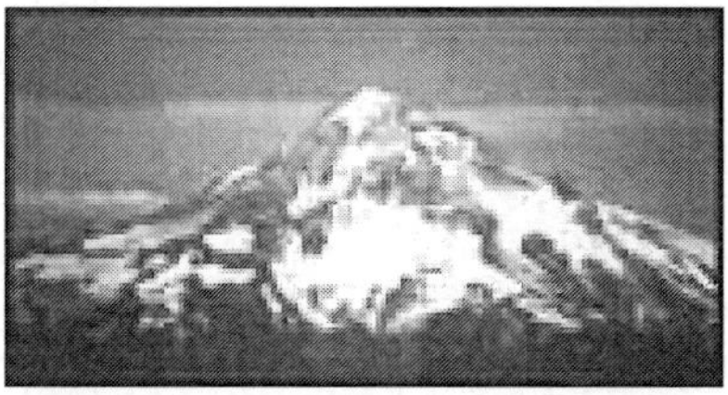

RADIATIVE FORCING AND CLIMATE SENSITIVITY

The radiative forcings from halocarbons and ozone to assess their roles in climate change. To make this comparison we have to assume that radiative forcings can be compared between different mechanisms and radiative forcing is related to the climate change issue of interest. The first assumption is equivalent to saying that the climate sensitivity is constant between different mechanisms and is implicit whenever direct and indirect radiative forcings are compared. However, several climate modelling studies have found that for many climate mechanisms ë varies with the latitude of the imposed forcing, and is higher for changes in the extratropics than for changes in the tropics. Additionally, several climate-modelling studies have compared the climate sensitivity for stratospheric ozone increases and generally found that global stratospheric ozone increases have a 20–80 per cent higher climate sensitivity than carbon dioxide.

This finding has been attributed to an additional positive feedback that results from an increase in stratospheric water vapour, which in turn arises from a warmer tropical tropopause. However, the latter studies all used idealized stratospheric ozone changes, and to date no study has performed similar analyses with realistic ozone changes. Climate models also typically have different responses to equivalent forcings from carbon dioxide and from other WMGHGs; these results indicate possible differences in climate sensitivity for CO_2 and halocarbons. The two studies that have examined this directly reached contradictory conclusions. Hansen *et al.* found halocarbons to have about a 20 per cent larger climate sensitivity than carbon dioxide, experiments. In contrast, Forster and Joshi found that, because halocar- bons preferentially heat the tropical tropopause region rather than the surface, climate sensitivity for halocarbon changes was 6% smaller than for carbon dioxide changes.

The trade-off or partial cancellation between direct and indirect radiative forcing only occurs in the global mean: the latitudinal forcing patterns actually complement each other. Both stratospheric ozone depletion and increases in halocarbons realise more positive forcing in the tropics and more negative forcing at higher latitudes. Therefore, the variation in climate sensitivity with latitude of any applied forcing would likely mean that, even if the halocarbon and ozone forcing cancelled each other out in the global mean, an overall global cooling would result. Further, the patterns of surface temperature response may well be distinct. Radiative forcing estimates are measured as changes from pre-industrial times and are indicative of the equilibrium surface temperature change one might have expected since then. Radiative forcing is not necessarily indicative of patterns of temperature change, transient temperature changes or other metrics. From any time period, such as the present, future temperature changes will come from the combination of temperatures continuing to respond to past radiative forcing and any new changes to the radiative forcing. The rate of change of radiative forcing is potentially more important for evaluating short-term climate change and possibilities for mitigation than the total radiative forcing. It is therefore useful to examine the rate, as well as the total radiative forcing.

In summary, adopting radiative forcing and examining its rate of change are among the best methods for comparing the climate roles of carbon dioxide, ozone-depleting gases, their substitutes and ozone. However, radiative forcing gives an estimate only of the equilibrium globally averaged surface temperature response, and could lead to errors of around 50 per cent when comparing the predicted global mean temperature change from ozone and the halocarbons with that from carbon dioxide.

Direct Radiative Forcing of Odss and their Substitutes

The direct radiative forcing from the ozone-depleting gases and their substitutes are relatively well known. The radiative forcing of the substitute gases is re-assessed in this report. The radiative forcings of the ODSs have individual uncertainties of about 10 per cent, and together they have contributed about 0.26 W m–2 to the total WMGHG radiative forcing since 1970. Past changes and a future scenario for the abundances of the ozone- depleting gases. A range of future scenarios for the substitute gases. In this part, we examine scenarios for future radiative forcing of the ODSs themselves. One possible future scenario is the A1B scenario from the Special Report on Emission Scenarios, which for the ODSs is consistent with the Ab scenario from WMO. However, because of their long lifetime in the atmosphere and their continued emission from the 'ODS bank', they could dominate the radiative forcing for the next four to five decades.

In this scenario HFCs dominate the radiative forcing by the end of the century, but this is only one example of several possible scenarios. As the production of ODSs has now essentially ceased, the principal issue for the emissions of ODSs concerns the treatment of the ODS bank. If emissions of ODSs from the bank continue at their current rate, ODS radiative forcing will still decrease. However, for at least the next two decades, the continued emission from the ODS bank is expected to have a comparable contribution to the total radiative forcing with that of HFC emissions. This calculation does not take into account the effect that ODSs have on the stratospheric ozone radiative forcing, which is discussed next.

Indirect Radiative Forcing of Odss

IPCC gives the value for the radiative forcing of stratospheric ozone as – 0.15 W m–2, with a range of ±0.1 W m–2. This radiative forcing primarily arises from mid-and high-latitude ozone depletion in the lower stratosphere. Past calculations of this radiative forcing have extensively employed observed trends in ozone. It is generally assumed that these observed ozone trends have been caused entirely by emissions of ODSs, so the negative radiative forcing of stratospheric ozone can be thought of as 'indirect'. For the ODSs it is possible to calculate EESC values and use them to scale the ozone radiative forcing. These values have been calculated by assuming that EESC values above a 1979 background level give a forcing that scales with the –0.15 W m–2 best-estimate forcing from IPCC over 1979–1997. This calculation follows the approaches of Daniel et al. and Forster and Joshi. It results in a value of the indirect forcing that is purely from stratospheric ozone loss and ignores any possible changes to tropospheric chemistry and climate. Uncertainties in the indirect radiative forcing of stratospheric ozone arise from many different factors:

- The quoted IPCC range of ±0.1 W m–2 for the radiative forcing of stratospheric ozone is based on differing model results and arises primarily from the underlying uncertainty in the vertical distribution of the ozone trend relative to the tropopause. The future radiative forcing from ozone is in fact more uncertain than the range quoted in IPCC because there is a lack of confidence in the details of future ozone changes, particularly near the tropopause, and many different scenarios are possible. Although most of the ozone changes can be attributed to ODSs, a sizeable fraction of the NH changes could have other causes, especially those that occur close to the tropopause. This means that a small but possibly significant part attributable to ODSs. The simple linear scaling between EESC values, above a 1979 background level, and the stratospheric ozone radiative forcing is used for obtaining an approximation to the ozone radiative forcing time-series. It was first proposed by Daniel et al. For EESC values below the 1979 level there is assumed to be no ozone depletion and no indirect radiative forcing.

❒

Index

A

Abundances 122, 191, 217, 235
Accelerated 13, 20, 55, 90, 190
Accumulated 17, 28, 151, 235
Assumptions 60, 216, 224

B

Barograph 170
Biodiversity 9, 16, 92, 100
Biome 30

C

Catastrophic 16, 46, 91
Chemistry 159, 194, 217
Circulation 36, 80, 114, 140, 225
Climatologist 221, 223

D

Deforestation 9, 47, 98, 189
Drastically 22

E

Ecosystem 89, 101, 197, 207
Estuaries 120, 229
Fascinating 26, 175

F

Fascinating 26, 175
Forest Management 199
Forecasting 27, 173, 185

G

Gaseous 20, 78, 136, 161
Glaciation 30, 187, 228
Global Climate 9, 45, 103, 187, 228

H

Hazardous 83, 84
Herbicides 209
Hurricane 23, 90, 172

I

Illustration 28, 204
Instabilities 192
Irradiance 103, 106

L

Legislation 119, 165

M

Manifestation 191, 216
Massachusetts 174, 182
Methylhalides 143

N

Nickel 166
Nutrient 191, 205, 222

O

Oceanographic 26, 133
Ozone 129,133, 143

P

Palaeoclimatic 190, 217
Peninsula 226, 229

R

Radiated 10, 63, 179
Ray-Pilots 176, 179, 180

S

Scatterometre 26
Sequestration 100, 211

T

Tetrachloride 132, 143, 144
Thermohaline 114, 189, 198
Troposphere 127, 133, 136, 172, 202

U

UARS 24, 149
Urbanization 14, 53, 121

V

Velocity 155, 175
Vernalization 209

W

Weather 121, 132, 142, 219, 226

Z

Zooplankton 111, 117

❑❑❑